Spatio-temporal Approaches

Series Editor
Anne Ruas

Spatio-temporal Approaches

Geographic Objects and Change Process

Hélène Mathian
Lena Sanders

iSTE

WILEY

First published 2014 in Great Britain and the United States by ISTE Ltd and John Wiley & Sons, Inc.

ISTE Ltd
27-37 St George's Road
London SW19 4EU
UK

www.iste.co.uk

John Wiley & Sons, Inc.
111 River Street
Hoboken, NJ 07030
USA

www.wiley.com

Library of Congress Control Number: 2014947782

British Library Cataloguing-in-Publication Data
A CIP record for this book is available from the British Library
ISSN 2051-2481 (Print)
ISSN 2051-249X (Online)
ISBN 978-1-84821-552-8

Contents

Introduction

The objective of this book is to introduce a constructed set of concepts, methods and approaches to represent and understand the evolution of social and environmental phenomena embedded in space. It relies on applications concerning not only human geography (field in which the authors have most practical experience), but also environmental geography and archeology in order to test the genericity of the proposed approaches. These three disciplinary fields are actually different on the thematic level, but they share similar epistemological and methodological stakes. Indeed, irrespective of the phenomenon (whether it is about monitoring glaciers in the context of climate change, about urban sprawl and the consequences of metropolitanization or about the transformations of the landscape), the aim is to consider how it is embedded in space and how it evolves over time. The ways of combining theme, space and time are discussed both in social and natural sciences, and it has given rise to numerous formalizations. In 1960s, Berry (pioneer of the "New Geography") proposed the concept of the "geographic information matrix" to apprehend these three-dimensions in geography [BER 64]. The "triad" (*what, where and when*), of geomatics inspiration, was developed by Peuquet in 1980s in the context of development of geographical information systems (GIS)[1] [PEU 84]. These approaches were then enriched by referring to the work of philosophers. Galton, for example, proposed the concept of "hyperobject" for an approach fully integrating space and time [GAL 04]. All these approaches will be combined here.

The book clearly falls within the scope of the field named "spatial analysis". Definitions about it are various, some highlight the more technical aspects, others

1 *Geographical information system* (GIS) refers to the computing environment integrating methods and tools for the process of geographical information, whereas *geographical information science* or geomatics refer to the set of concepts and methods underlying the processing of geographical information.

insist on the methodological aspects and others finally propose a more comprehensive scientific vision. Our positioning is located at the interface of two definitions: – for the first, spatial analysis is "the formalized study of the configuration and the properties of the produced and lived space of human societies" [PUM 97]; – in the second, spatial analysis is a "set of techniques and models that apply formal structures, generally quantitative, to systems in which the main variable evolves significantly through space" [LON 96]. These two definitions put forward the principle of a formal approach, either at the conceptual, observation or measurement level. Our approach relies on a systematic coming and going between these two definitions, each one supporting the other. We will multiply the points of views about each concept and each method, comparing, for example, the points of view of the philosopher and computer scientist on questions of ontology, those of the geomatician and philosopher on spatiotemporal concepts, those of the statistician and simulation modeler to explain the evolution of societies, of their spaces and environments. Thus, the objective is to articulate concepts from information science, complex systems theory and thematic fields. The latter concern not only geography but also archeology that offers, through the temporalities involved, very stimulating stakes to reflect upon the spatiotemporal approaches.

The models occupy a central place in spatial analysis. Rather than deepening the technical aspects that are related to them, the choice has been made to clarify and discuss different modeling approaches and to emphasize the multiplicity of points of view. "Model" is a polysemic term, and misunderstandings may emerge when specialists in different areas collaborate, notably as far as we are concerned here, computer scientists, geomaticians, geographers and archaeologists. Two definitions, here again, very well reflect, in their complementarity, the position adopted in this book:

– "The model is a schematic representation of reality, developed with the purpose of understanding reality and explaining it" (Haggett [HAG 65] and Durand-Dastès [DUR 92]);

– "To an observer B, an object A* is a model of an object A to the extent that B can use A* to answer questions that interest him about A" (Minsky, [MIN 65]).

While assuming a solid anchorage into reality, an aspect stressed in the first definition, it also seems essential to adopt the distantiation from reality that the second definition implies. This distantiation allows, as a matter of fact, reasoning not only on what may have existed, but also on what could have happened or could eventually happen in the future. In both cases, the model is a scientific support tool for reflection.

The risk of misunderstanding with regard to the term "model" is accentuated by the fact that, faced with a given problematic, the models that are mobilized at different stages of the research are of a different nature: conceptual model, data model, statistical model and computer simulation model. Rather than confining ourselves to a particular point of view, our objective is precisely to address the concept of models in a broad manner: models to build objects, to generate, manage and describe information, models to explain and understand spatial phenomena and their evolutions. The emphasis will be on the conceptual framework rather than on the technical and operational aspects that can easily be deepened elsewhere. In order to explore and analyze relationships in space and time, we will favor the application of simple, classic, robust statistical methods, of common use and accessibility in social science, by showing how certain couplings or sequences allow addressing complex issues. When this is necessary relative to the problematics, especially when it comes to understanding how structures emerge from the interactions between elementary entities, the methods coming from complex systems theory will be called into action. The privileged modeling methods from this domain, such as cellular automata and multiagent systems, which allow simulating the emergence of spatial structures from the interactions between entities considered as elementary, will be, therefore, presented.

Space will be apprehended in a multiscale manner, with a thorough reflection on the meaning of the entities that make it up, and privileging a systemic perspective, with particular attention paid to the interactions operating in space according to different temporalities. The manner in which the concept of system is called upon in spatial analysis has become richer over the past decades. In its simplest definition, a system is a set of entities that interact and hence form, a whole, which has limits, and which is distinguishable from its environment. The expression "the whole is more than the sum of the parts" is classically associated with this vision. Thus, a spatial system is a system composed of a set of spatial entities interacting, for example, system of cities, settlement system, system of land use and landscape system. For over a decade now, human and social sciences have been focusing on the field of complex systems that offer the advantage of taking into account several levels of entities: "complex systems consist of qualitatively different organizational strata, in particular, a microlevel and a macrolevel. Between them there exist a "bottom up" and "top down" interaction, i.e. a quasi-cyclical causal relation" [WEI 06]. Spatial analysis is further concerned with a whole set of intermediary levels which qualify as meso-geographical. The diversity of these levels and of the entities that are associated with them is today renewed through the proliferation of fine-grained data allowing for a great variety of aggregations. In addition to the effects from nesting different levels, Dauphine and Provitolo [DAU 13] identify two other sources of complexity. On the one hand, a number of simple systems on the formal plan may show erratic and unpredictable behaviors (properties that have significantly mobilized mathematicians and physicists from complex systems theory). On the other hand, in a multidimensional context, the significant number of domains interacting represents an additional source of complexity, especially when environmental and human systems interact.

This systemic design implies identifying and clarifying the multiple interactions between the entities of a same level (horizontal interactions) and between the levels that constitute the system (vertical interactions). To use the expression of Irwin and Geoghegan [IRW 01], our position is closer to the "creative" use of spatiotemporal data and further from the question of the "correct" use. The latter has given rise to a large number of specific developments to respond to essential statistics issues such as the correction of the effects of spatial dependency or of process heterogeneity in space [GRI 91, ANS 95, DUB 14] etc. Actually, geographers, historians and archaeologists have strong assumptions on the role that space plays or may play in observed phenomena. We have placed the emphasis on how we can take it into account in formalizing space, as well as time, and integrating them in the analyses. The most common way of taking space into account is without any doubt through the distance to a structuring element. The challenge lies in moving beyond this single linear formalization of space and introducing either multiscale effects (for example, through the multiple and overlapping influences of centers of different levels according to their distance), or other formalizations such as neighborhoods, territorial contexts and even going to the point of integrating "fuzzy" belongings. Space is no longer only analyzed to understand similarities but also to extract differences [FOT 00]. We are committed to building information and also integrating results of models based on assumptions (for example, the potential model) to bring original and useful insight.

The pedagogical aspects will not be overlooked, but it seemed to us essential to present applications that were "life-size" and not overly simplified. The choice was, therefore, made to privilege examples derived from research practice, considered in all their complexity and richness. In practice, the spatiotemporal problematics contains, in general, two sources of complexity: – the objects of interest themselves, often of multiscale and evolving character, that have to be built in order to follow them (identity and change); – the approaches adopted that may be hybrid, or include couplings, or sequences of methods. A crossbreeding of these two aspects is necessary to describe complex phenomena and to understand the functioning of spatial systems such as settlement systems, educational systems, socio-environmental relationships, river systems, etc. The objective is to propose a generic framework to conduct such an approach in the context of social sciences where each analysis is not an end in itself ("flawless piece of result"), but a useful brick in the construction of the knowledge of the phenomenon analyzed [FOT 00].

The book is organized into four chapters (Figure I.1).

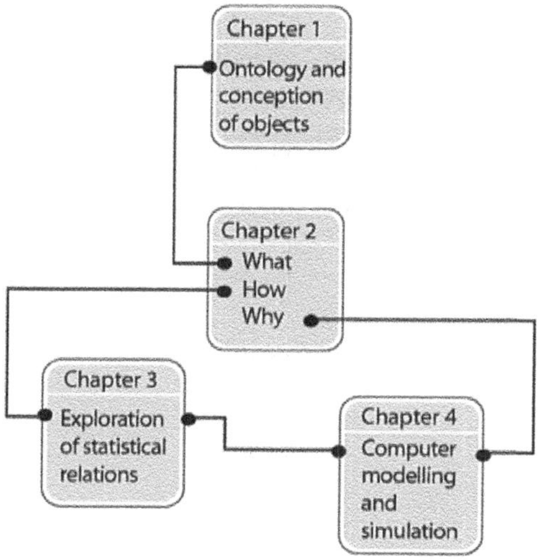

Figure I.1. *Book structure and relationships between chapters*

Chapter 1 is devoted to the design and construction of objects, attributes, relationships and processes that are associated with the problematics raised. We defend the idea that the quality of the formalization upstream, irrespective of the degree of sophistication of further processing applied, is a necessary condition for ensuring the relevance of the results. With the enlargement of the observation fields (fineness of temporal and spatial granularity and multiplication of observation means), the reflection on the meaning of the entities of interest has to be intensified.

Chapter 2, entitled "From Empirical Questioning to Spatio-Temporal Modeling", is dedicated to the transition, more delicate than most often thought *a priori*, between the thematic questions formulated by the researcher or practitioner, the adequate methods to answer them and the observable entities available. In this chapter, we clarify the potential gaps between these three sets, and we exemplify the sequence of methods related to specific issues. This chapter represents a connection between Chapter 1 and Chapters 3 and 4, in the sense that it enlightens the categories discussed from a theoretical point of view in Chapter 1 and introduces through practical examples the problematics that are then deepened in the following two chapters.

Chapters 3 and 4 address the practices developed to present and understand the change of spatial systems, both at the level of the elementary entities that make them up and that of the systems-entities themselves. Chapter 3 is dedicated to the description and statistical explanation, the ways of studying *how* the interrelationships between different phenomena embedded in space evolve. Chapter 4 deals directly with the processes at stake during change and the modeling of the mechanisms underlying these processes. Particular emphasis is put on computer simulation methods. Figure I.1 illustrates the position of these two chapters in direct relation with the first two and how they complement one another.

1

Building Objects in Time

The world is composed of "things" that we conceptualize as objects "with the purpose of building knowledge from it" [DEB 04]. Information is increasingly more abundant. It is also available at more diverse granularity levels due to technological progress in the field of acquisition and storage. In such a context, the possibilities of observation are multiplying for the researcher. From this multiplication of possibilities arises the need to *choose*, and especially that of clarifying the choices made: reflection first on the objects that we consider as relevant relative to the problems posed (conceptual dimension); choice then of the observable entities that will allow us to study these objects of interest (empirical dimension); choice finally of what we will observe about the characteristics and behaviors of these objects (heuristic dimension). Therefore, the purpose is to build objects from observable "things" in the empirical world, and to give them a meaning relative to the problematics at stake. This path is not always as immediate as we would like. Two concrete examples can illustrate it effectively :

EXAMPLE 1.1.– Let us suppose that the decision to build a commercial establishment X in a city A depends on the growth of the population in this city during the last decade. The decision maker requests three consultancy firms to estimate the evolution of the population of city A. The first concludes with a decline, the second with a stagnation and the third with a growth. The decision maker is then perplexed, wondering about the respective competencies of the three consultancies, and in total uncertainty as his/her decision. A more thorough examination of the work carried out showed that each of the three consultancies had adopted a different definition of what a city *is*: – the first consultancy firm based its measure on a political–administrative criterion and used the evolution of the population of *municipality* A; – the second used population data corresponding to the *urban unit* A as the French National Institute for Statistics and Economic Studies (INSEE) defines it, that is an entity made up of a central commune and neighboring communes, according to a morphological[1] criterion; – finally, the third used the population statistics of the "*urban area*", an entity constructed by the INSEE according to a functional criterion. According to this criterion, a city incorporates the totality of the communes of which part of the active population works in the urban unit. The conceptual object on which the decision maker poses its questioning is, therefore, *the* city A of which he/she wishes to know the population evolution. In the empirical domain, each of the three "observables" (all three supplied by INSEE, but each corresponding to a different delimitation of city A) referred to in this example is *a priori* relevant to answer the question. The ambiguity comes from the fact that the city concept has not been sufficiently specified. Depending on the nature of the establishment X and the targeted customers (the inhabitants of the central commune of A or the active population working in the urban area of A), one of these "observables" will be more appropriate than the others.

EXAMPLE 1.2.– in the fields of geosciences and environment, this type of question arises with the same sharpness. In a similar way to the previous example, let us raise the question about monitoring the evolution of glaciers. Indeed, it is essential in order to evaluate the associated risks. The dynamics of glaciers is an indicator of climate fluctuations. It is associated with a number of events that are sources of risk for the surrounding human activities: avalanches, flash floods, inundations and mudslides. In addition to these disasters, a number of consequences also affect economical activities such as that of mountain tourism. Even if the expert agrees on the conceptual definition of a glacier, irrespective of his disciplinary training, glaciologist, geomorphologist or geophysicist, the specialist does not use the same approach to analyze its dynamics and displacement. The points of view will differ both in the choice of information sources and in the methodology for dynamics monitoring: some will favor the monitoring of field measurements as that of glaciological beacons; others will use high-definition satellite images to delineate the glacier at different dates and observe the evolution of the surface; get others choose not to start from the delimitation of the glacier, but to build an indicator whose monitoring in time provides information about the movement of the glacier, for example, the glacier equilibrium line altitude (ELA)[2] [COS 11]. All these approaches allow apprehending the same question (the dynamics of the glacier), but the quantitative results will be different, since they correspond to different measures, each appropriate to a particular scale, the fine-scale of the glacier for the glaciologist, and more often the regional scale for the geophysicist.

The first example illustrates how an insufficient specification of the city entity leads to the observation of different empirical objects (the commune, the "urban

1 In this framework, the morphology of a city is defined from the built up area. A morphological criterion is then based on the continuity of the built up area.

2 "The glacier ELA refers to the altitude of the theoretical line between the zone of glacial accumulation and the zone of glacial ablation" [COS 11].

unit" and the "urban area" depending on the case), which lead to conflicting results. The second example illustrates the possibility of a variety of points of view on the same phenomenon. In both cases, the challenge consists of choosing the scales and observable empirical objects that are the most relevant relative to the questions asked. The keywords of this chapter are thus *objects* on the one hand, and *construction, representation* and *change* (of/in these objects) on the other hand. The sense that will be given to these concepts throughout the book, with the objective of modeling phenomena embedded in space and happening in time, has to be defined. Several points of view are to be taken into account, each with its own specificities. That of the domain expert[3] (geographer and archaeologist in the context of this book) differs from that of the formal sciences (computer scientist, geomatician[4] as well as philosophers). The objective of the former is to represent, describe and understand a social or environmental phenomenon, for example, the strengthening of educational inequalities, the growth differentials between cities, the dynamics of a glacier, the changes in soil occupation, the evolution of the interaction scope of a city during the middle ages or the practices and rhythms of individuals' mobility in different spatial contexts (for example, a train station or a touristic place). The objective of the latter is to build generic representation or modeling media, independent of the types of questions and the studied objects. A geographer and a geomatician will, therefore, have different points of view on geographic objects: for example, in an analysis of the public space, a street may be defined as "a system of places close to each other, connected through practices" [FLE 07]. This design just overthrows the representation that is generally made in geographical information systems where the street is most often represented by a line connecting places.

In addition, faced with the same empirical question, the points of view of domain experts from different areas, for example, archaeologists and geographers, will also differ, notably on how to apprehend time and space. In parallel, at the heart of the formal sciences, including philosophy and information technology, the ways to specify objects and processes at stake differ.

This diversity of points of view, rather than being a source of misunderstandings, can be considered as an asset to the extent where the research for consistency that it requires, constitutes in itself a step forward in the reasoning of the domain expert. Furthermore, the precise conceptualization of the object of interest is necessary to obtain an interpretable formal description that can be implemented on a computer

3 We will call "domain expert" the specialist of a field in the social sciences, raising questions about a given thematic and having an expertise about the domain considered [LIV 10]. The examples raised in this text will mainly fall under the fields of geography and archeology.

4 We will call geomatician the specialist in the science of geographic information, who raises issues of formalization incorporating geographic reasoning. He/she is a specialist of computer developments structured for the acquisition and implementation of geographical databases, for information processing in dedicated systems and for the representation of this information.

system with the least possible ambiguity [PHA 14]. This chapter focuses on two large questions: (1) on the identification, construction and categorization of the objects associated with the posed question; (2) on how to apprehend change, either at the level of the objects themselves or at that of the attributes that characterize them, of the relationships among them, of the processes that underlie them. The categories of objects will be discussed from an ontological approach situated at the interfaces of philosophy, computer science and geomatics (section 1.1). We will then discuss the different ways of dealing with change of objects in time (section 1.2). The main objective is to show the interest of an approach that allows moving from a sociospatial, historical or environmental problematics to a conceptualization in terms of objects. Most considerations are anchored in the operational area of management and analysis of geographical information, but the objective is not to address the operationalization of these designs in formal representation languages.

1.1. Different points of view on ontology

The first step consists of a start from an ontological point of view to discuss "things" at stake during the description and modeling of a spatio-temporal question, whether it is a social or environmental one. Ontology, the study of "the being as being" according to Aristotle, must, therefore, allow specifying "things" that we wish to study, whether from the conceptual or the empirical point of view (whether database, statistics or simulation are considered). Before giving the definition that will be adopted in this book, we conduct a quick review of some definitions put forward in the fields of philosophy, information technology and information sciences, stressing the specificities that belong to each one of them. Different categorizations that can be made about the "things" that we are studying are then discussed, and an example of object construction concludes the first part of this chapter.

1.1.1. *Defining ontology*

Smith, a philosopher of Aristotelian inspiration, suggested the following definition of ontology: "the science of what is, of the types and objects structures, properties, events, processes and relationships in every reality domain .../... of what *could* exist" [SMI 03]. The use of this conditional, referring to "things" that have not necessarily been observed, is essential when we place ourselves in a modeling and simulation perspective involving artificial worlds. In computer science, Gruber defines ontology as a "specification of the conceptualization of a given domain" [GRU 93]. This definition is consistent with the previous one in the sense that this "specification" consists of clarifying the objects, properties and relationships mentioned in Smith's definition. While the latter refers to world "things" ("in every reality domain"), Gruber's definition falls under the framework of knowledge-based systems. What *is*, what *exists,* relates therefore to what can be *represented* [GRU 93].

In the field of information systems, Chen [CHE 76] relies on an ontology made up of three fundamental elements, entities, attributes and relationships in order to develop what he calls an "entity-relationship modeling". For him, the entities designate "things" that are identifiable, distinguishable from their environment and that correspond to "objects" in Smith's terminology. In order to illustrate the different terms of this ontology, let us consider the example of the school domain in which students and schools represent two types of "objects-entities". The attributes describe the characteristics of these "objects-entities" (for example, for junior high schools, the public/private status, the results from the French "brevet" national exam). These attributes correspond to Smith's "properties". The relationships concern on the one hand the links between the "objects-entities" and their attributes (schools are, for example, public or private, have such number of pupils, etc.), on the other hand between "objects-entities" of different nature (such student attends such school), or even between different attributes (the results of the French "brevet" are better in such type of school) and finally the links between "objects-entities" of the same nature (exchanges between same class students, proximity between schools or flow of pupils changing schools).

These different points of view formulated, respectively, by philosophers, computer scientists and information systems specialists, converge in their ambition to describe the world/a world using a generic conceptualization, but it is interesting to point out the nuances in their approaches. In this way, differences of opinion exist between philosophers and computer scientists about the ontologies. The fact that the objects do not necessarily constitute the favored input for philosophers when it is often the case in computer science is an example. Indeed, for philosophers, processes, for example, could replace them [LIV 09]. Smith [SMI 98] for his part, clarifies the difference in point of view between philosophers and "information scientists" by distinguishing a "reality-based" ontology that has an objective to describe the world in its "reality" and an ontology that he qualifies as epistemological and that is associated with a particular conceptualization of the world (among others). In addition, Peuquet [PEU 02] points out that Chen [CHE 76] presents his three fundamental concepts and their articulation at the center of his entity-relationship model without making reference to the philosophical literature on ontologies. This approach was then developed in a progressive and autonomous manner in the field of computer science to build a "theory of database model design". This perspective of the *model design,* representative of computer science, implies a *bottom-up* approach, while the philosopher falls instead under a *top-down* perspective by seeking to identify the most general categories possible to respond to the largest range of questions possible [LIV 14].

The field of geographic information science ("GIScience") is more recent and benefits from the maturation of these ontology issues, allowing us to take account at the same time of the aspects developed in philosophy and in computer science, by seeking to integrate the advantages of each point of view. From a computer science perspective (*bottom-up*), the objective of implementation involves constraints that may cause excessive simplification of the higher level abstractions that tend to be favored by the philosopher. The risk is then to reduce the interest and the scope of the elaborated ontology not taking into account, for example, its evolutionary character [PEU 02]. Conversely, if we systematically privilege the principle of generality, the risk is that the ontological framework is too general to allow the domain expert to operationalize it to answer his questions. Consequently, the challenge consists of finding a middle ground between these two risks.

Further in this book, we will adopt the definition proposed by Livet [LIV 10] that is situated at the intersection of these different approaches: the ontology consists of analyzing a domain, by identifying the *relevant entities* (objects, properties, relationships, events and processes), and *the operations* that can be carried out on these entities. We will explain what are these entities and operations with regard to a geographical problematics:

– The *entities* are of five types: (1) the *objects*[5], whether they are geographical objects such as rivers, roads, plots or spatial units such as municipalities and cities or even localized objects such as individuals, households and dwellings; (2) the *properties* that characterize these objects; they refer in the empirical practice to the "attributes", which may be of a quantitative nature (length, surface area or number of inhabitants) or qualitative nature (type of land use or political color of the municipality); (3) the *relationships* between these entities, which relate, in the sense described above [CHE 76], different attributes between themselves, the objects and their attributes or even different objects between them, for example, the proximity relationships (inclusion, adjacency and distance) and exchanges (migratory flows and information flow); (4) the *events* that characterize the appearances, disappearances and abrupt changes of these same entities; (5) the *processes* that refer to what cause these entities (objects, properties and relationships) to change over time.

5 There is no consensus in the literature on the difference between entity and object that are sometimes considered as synonyms. This is not the case in the definition adopted in this book where the entity refers either to objects (human beings, spatial units such as communes or towns), properties, relationships or processes. To avoid any confusion further in the text, we will use the term entity for this generic sense and we will use that of spatial unit to designate an administrative or spatial entity.

– The *operations* related to these entities allow modeling their structure, functioning and evolution. We can distinguish three main families of operations: (1) the measurements that allow specifying the properties of objects (for example, size and density measurement); (2) the functions that allow characterizing the relationships between objects, whether they concern proximity (in terms of similarity or genealogy) or interactions (for example, trade, hydrologic flows and access time); (3) the rules and functions that allow linking an event to a change or generating a series of successive changes. In a model on the growth of cities, for example, the evolution of a city population may consequently be formalized from a differential equation (logistics equation, for example, expressing an exponential kind of growth when the population is far from the carrying capacity of the city and slowing down as it approaches it) or from a rule that can be formulated in the following way: "if the potential of interaction with other cities is of a given intensity, then the growth rate of the population is so much".

This categorization has proved fruitful when frequent moves to and fro are made between the empirical question, the associated information system and the modeling. Faced with the same empirical question, multiple ontological choices are possible as a result. For example, let us suppose that we are concerned with the evolution of a settlement system in the long term and that we want in particular to model the change from villages to cities. The following two ontologies are possible:

– first case: we consider the existence of two types of objects, cities and villages, which are then apprehended as two different "things"; each one will be characterized by *properties*, and we will be able to define *relations* between these two types of objects, for example, relationships of functional dependency, as well as *operations*, for example in the form of a transformation rule from one type of object to the other;

– second case: a single object, the settlement unit, is considered, and the distinction city/village is then understood through the properties of this object: it can be a property captured directly by a simple attribute with two modalities or several properties related to quantitative attributes such as the number of inhabitants, the economic profile, the level of services or the range of exchanges. In the latter case, the purpose will be to define the operations, usually from recurring rules to thresholds, which will allow characterizing the city or village property of the settlement unit.

The researcher will choose one or the other from these ontologies on the basis of his assumptions on the differences between the objects *city* and *village* and on the possibilities of transformation from one to the other. If the assumption is that there is a semantic difference between city and village, with properties and relations associated with the village being qualitatively different from those associated with the city, the first case is to be considered. This is, for example, the point of view of

Garmy [GAR 12] in his work on settlement systems during the Roman era. The variety of their profiles and functioning modes thus leads him to characterize the settlement units of the Languedoc according to a precise functional typology that cannot be reduced to a simple city/village dichotomy. However, the second case is adequate for the hypothesis of a semantic continuity between these two types of objects, i.e. if it is assumed that the objects are of the same nature and characterizable by the same types of properties and relationships. Simple differences in the properties (for example, the hierarchical level measured by the size of the population or the level of the existing trade) are enough then to account for their differences. This is often the case when geographers or economists aim to model the emergence of cities and systems of cities from an initial situation in which only agricultural villages with few differences existed [AXT 06, BAT 01, SAN 97, SCH 13].

1.1.2. *Qualification of the objects from an ontological perspective: "bona fide" versus "fiat" objects*

Facing the question of falling under the scope of spatial analysis, the challenge is to ensure consistency between the entities of interest in "reality", those that are observable in the empirical world, and those that are appropriate to introduce in the information system that is implemented to respond to the questions raised. These entities can be *simple* (an individual, a road, a parking or a river) or *composite* (a household, a subdivision, a city or a hydrographic system). This distinction between a *simple* or a *composite* object depends on the thematic domain or may be discussed. The composite object is composed of simple objects, themselves having a meaning relative to the question asked (note that from the theoretical point of view, it would be always possible to decompose down to the atom; the position adopted here consists of adapting the level of decomposition to the objective and in discussing the choice made). The information allowing building these entities often originates from various sources.

Any object mobilized during a research is a *built* object in the sense that it is conceptualized to meet an objective. Some of these objects are close to the common sense, in the sense that they are concepts used in everyday life, corresponding in general to physical entities, for example, individuals, buildings, streets, rivers and lakes [PEU 98], whereas others correspond to more abstract concepts referring to more elaborated constructions. Households, cities and countries are some examples. We propose to start with the distinction proposed by the philosophers Smith and

Varzi [SMI 00], between *bona fide*[6] objects and *fiat*[7] objects, by adapting them to the objects encountered during spatial analysis works:

– *bona fide* objects correspond to "natural" objects of common sense [SMI 00]. They have a material and physical anchorage, and the authors evoke "genuine objects", i.e. "real", "authentic" objects. These objects exist and are distinguished from their environment regardless of what Galton named "human partitioning activity" [GAL 03]. They correspond, therefore, to a "reality" that can be described as objective [PEU 98];

– *fiat* objects are constructed objects, which exist according to an administrative, social or political convention [SMI 00], or in the imaginary, as a concept [HOR 00]. These objects are, therefore, derived from a human partitioning activity, and it is useful to distinguish two cases: – the objects corresponding to a political or administrative convention, for example, such is the case of countries and administrative regions; – the objects constructed by "expertise", with a scientific or operational objective. The dichotomy proposed by Terrier [TER 98] between "power zoning" and "knowledge zoning" illustrates this distinction from the point of view of domain expert well. The first term refers to a splitting of the space in zones corresponding to the exercise of a power (typically the municipalities and departments). The second corresponds to a splitting of the space made by institutions and researchers with the purpose of answering a question (for example, the mobilization of census tracts to study at what rate urban sprawl increased).

Of course, the distinction between *bona fide* and *fiat* categories is not always immediate and intermediate cases can appear. A forest, for example, can be viewed from both angles: it is distinguishable in the landscape, but its delimitation is not always unequivocal. As an island, the group formed by the space bringing together the two departments of Corsica is of the first type; as an administrative unit, each of these departments taken separately corresponds to the second. The benefit in explicitly discussing the ontological nature of the object that is being worked upon is in first place epistemological. This reflection helps to ensure that there is no ambiguity in the definition of the objects, to control any potential gap between the object as it was designed and the object as the empirical approach allows observing and analyzing it, and finally to give meaning to the results of any processing carried out. We think it is also clarifying to combine this categorization corresponding to the point of view of the philosophers, with the methodological categorization on the

6 The Latin phrase "bona fide" meaning "in good faith" is used to characterize what is authentic in English as well as in French texts.

7 "*Fiat Lux*" is a Latin phrase present at the beginning of the Genesis. It is the first word of God, order given when he creates light during the creation of the world, translatable into English by "*let there be light*". This formula that experiences a great success since the 18th Century is used nowadays to mention slightly pompously an invention or a discovery." (Wikipedia).

simple or composite character of the objects at stake. This combination defines four cases (Table 1.1) that establish a first classification of the objects that are manipulated during spatial analysis. The examples are presented for information purposes, and some positions in the table could be questioned, such as the case of the forest (mentioned above) or even of the river. As a matter of fact, the latter may be viewed as a simple object making up a whole, elementary entity of the landscape, or as a composite object, composed of a sequence of river stretches delimited by some relevant features of the landscape (for example, change of direction).

		Methodological point of view	
Ontological point of view		Simple	Composite
	"bona fide"	Individual being, housing, building, tree, river...	Family, copse, river, forest, flock...
	"fiat"	Municipality, country, farm, university...	Household, city, forest, built area...

Table 1.1. *Combination of two categorizations of objects with a few examples of each case*

1.1.3. *Specification of ontologies in the field of spatial analysis and geographical sciences: objects versus fields*

The ontological conceptualization will help to impose consistency between conceptual objects responding to the problematics posed and empirical objects, observable in the reality of the field. This step is essential in the path that leads from the facts to be described and understood to the data structuration to be mobilized to achieve this. The conceptualization and observation of empirical phenomena can be done in different ways. We will discuss in particular the dichotomy between the ontologies in terms of *objects* (discrete perspective) and in terms of *fields* (continuous perspective), opposition that also flows through other sciences, notably physics.

When focusing on spatial phenomena and performing spatial analysis, actually a first distinction is to be made between an ontology of the *discrete* and an ontology of the *continuous*. In the geographic information sciences, the first refers to an approach "object-based" and the second refers to an approach "field-based". In this domain, the dichotomy object/field (object based/field based) is a matter related to data and data models developed to store, manage and represent geographical information. Couclelis [COU 92] establishes a parallel between this dichotomy and the discussions conducted within physics on the opposition between the theory of elementary particles (making reference to the atomic and quantum properties of bodies) and a continuous field approach, also called plenum, supposing the continuity of space and time. This paradigm is useful to define a conceptual framework, and here we are recalling it to distinguish between two points of view on geographical space:

– the "discrete" or "object-oriented" point of view is more easily conceptualizable because it is well-anchored in common language [GAL 03]: individual beings, buildings and roads are objects of common sense, relatively easy to identify in the sense that they are "distinguishable" from their environment and associated with an identity. Once built, objects of the "fiat" type, such as cities or countries, share these properties and can be conceptualized by an object approach, of the atomistic type. The totality of these objects exists independently from their attributes, and they retain their identity when manipulated;

– the "continuous" or "field-oriented" point of view is often the work of geoscience or environmental science specialists. As a matter of fact, such approach is appropriate for geographic information that is presented in continuous form, for example, about land use or temperature. Space can then be conceptualized as a set of localizations to which attributes are attached ("a spatial field is a mapping of spatial locations to values", [GAL 04]). The nature of the information associated with localizations can be of any kind, qualitative (land use) as well as quantitative (for example, temperature and hydrometry degree). The most important thing is that the *field* perspective refers to a coverage of the space.

The adoption of one of these two points of view rather than the other depends either on the *a priori* position of the researcher or on his choice in front of a specific objective in which he considers one of the approaches more appropriate than the other. Most of the spatial phenomena can effectively be conceptualized from either an "object" or a "field" perspective: thus, a vegetation cover can be apprehended as consisting of objects, such as prairies or woods at the landscape scale, or of trees and shrubs at a finer scale or as a continuous field covering a given space [PEU 88, PEU 98]. Another example is forest fire that can be formalized as an object moving in space, or as a continuous field of fire intensity. The two perspectives are associated with different conceptions of space. In one case, the object is autonomous, it has its own attributes and space is a support, a referential frame in which the objects are positioned: "imposed upon" underlying fields according to Galton ([GAL 03]). The objects become to some extent the inhabitants of a space that would be otherwise empty. In the other case, the field forms a "plenum", where each localization is attached to a property: we are then faced with what Couclelis [COU 92] called a "continuous fabric". In practice, it is possible to extract objects from a "field-based" approach: the objects are carved out from the properties attached to localizations. These objects can then be characterized in turn by the appropriate attributes. Thus, Plewe [PLE 98] gives the example of the hill. It may be represented as a field of altitudes and is not delimited by a sharp border in nature; there is a gradation. On the other hand, if we want to represent it as an object, one is led to determine an altitude threshold and then delimit the hill with a clear and sharp line. Galton considers that the objects constructed following such an approach correspond to a higher level of abstraction than those related to a direct "object" approach. From the cognitive point of view, an observer can move from a plenum view, consisting, for example, of watching from a train the spatial organization of a landscape that is passing by to an object view where he identifies relative localizations (for example, the tree next to the church) or concentrations (for example, several juxtaposed silos) as illustrated in Figure 1.1. From the point of view of the computer system, it is, however, not easy to manage such round-trips instantaneously [PEU 02].

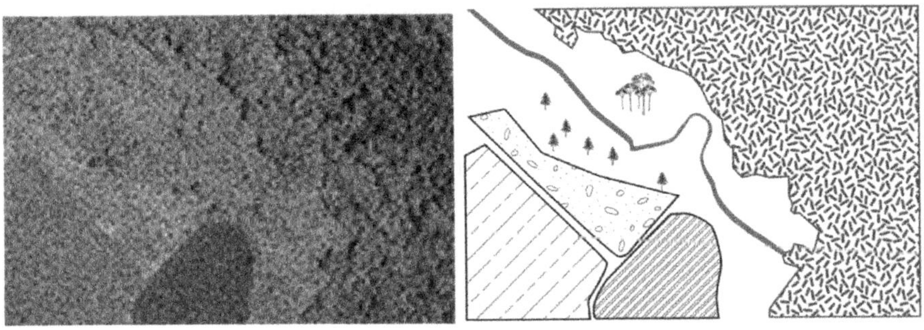

Figure 1.1. *Field-based and object-based approaches: two points of view on the same observable (source: UMR5600 EVS-ISG – ENSL K. Michell). For a color version of the figure, see www.iste.co.uk/mathian/spatiotemporal.zip*

These two perspectives are considered as dual, and some phenomena can be described according to each of these perspectives [PEU 02]. This depends on the phenomenon, of course, and on the scale at which it is considered: a tree will naturally be conceptualized as a discrete object and the temperature in field form. In contrast, the choice is to be made for vegetation coverage, and a number of other phenomena call for further discussion. As a result, Plewe [PLE 98] proposes a categorization into four conceptual models of space: – (1) the plenum is something that fills space, and the properties that matter can be measured at each point (for example, temperature, the nitrogen content of the soil etc.); – (2) The *regions of the space* delimited from the data that the author calls "*categorical coverage*"; it means, for example, built-up areas, green spaces, water surfaces etc. – (3) The *regions of the space delimited in a formal manner*, and not necessarily from the properties of the space. Such is the case of countries, or even of the whole set of census tracts making up an urban unit in official statistics definitions; – (4). The *object perspective* finally, in which the objects are thought as existing by themselves. Such is the case of the city of London, for example, which can be apprehended without reference to the way in which its delimitation has been established. Therefore, Plewe introduces two intermediate models that allow nuancing the fields/object dichotomy, but nonetheless a same phenomenon can fall under different categories depending on the point of view adopted.

Brogaard [BRO 98] for her part uses granularity as a filter to distinguish between two types of fields, one refers to a complete continuity while other refers to objects at a certain level of scale. She illustrates this latter case with the example of population densities. A density field is extrapolated from regions sufficiently thin for the density to be either 0 or 1, depending on whether an individual is present or not on these portions of space. Thus, the existence of this field is subject to the existence of individuals who serve as the basis for the measurement of the density, and which are “material objects”. She opposes this type of field to the one where the subject of interest is the salt concentration of a lake: there is only one portion of space where this concentration has a meaning, that of the lake, and the lake as well as the distribution of the salt in the lake are of a field kind. Brogaard proposes consequently to distinguish two types of geographic fields: – the fields that depend on the objects located in this field (“object fields”), the example being that of the population density; – the fields that depend on fields (“continuity fields”), the example being that of the salt concentration.

In general, we can also distinguish two types of objects: – *fixed* objects, whether they are clearly distinguishable from their environment (a tree or a building) or whether they are part of it (example of the hill above, or even of a river); – *mobile* objects, which move in space (for example, pedestrians, automobiles and ships), space playing then the role of support. Here again, going from an object type to the other can be a question of observation granularity. The spatial envelope of a demonstration, for example, may, therefore, be captured as a fixed object from an aerial photo, and this envelope can have fixed contours while the demonstrators, mobile objects, move in all directions.

1.1.4. *An example of empirical objects’ construction: the case of cities*

Here, we propose to illustrate the different stages that are associated with the passage from the conceptualization of objects to their concrete construction (this passage corresponds to the operationalization), mobilizing observables in the framework of databases. The “city” is a good example, and it even constitutes a quasi-iconic case in geography, of a built object[8] [BRE 13]. The passage from the conceptual object to the empirical object is not generally done directly and needs giving account of the construction process.

8 This example is drawn from multiple experiments led by Anne Bretagnolle: ANR Harmoni-cities http://www.parisgeo.cnrs.fr/spip.php?article38&lang=en) and ESPON DB program – urban data (http://www.espon.eu/main/Menu_ToolsandMaps/ESPON2013Database/).

Conceptually defining the city as a place of concentration of inhabitants, jobs, services, etc. allows making a mental representation of it. The aim is then to formalize this representation in terms of observables, and the path is still a long one before having made the whole construction process explicit that then leads to a representation in computing terms. This process is based on the categories presented above. Among the possible choices, a city can be defined as a political entity (municipality, metropolitan region, agglomeration community...), or as a morphological entity (dense space and/or built in a continuous manner), or even as a functional entity (employment center, area drawn by daily commuters...). If we rely on the first design of the city, there are only few questions to ask: indeed, the city in this case is an object of the *fiat* type, defined by convention or politico-administrative decision. In the case of the morphological definition, the first question that arises concerns without doubt the conceptualization of space: can we apprehend the city and its dynamics through the bias of a density field (Figure 1.2), where the city appears as a "spatial singularity", distinguishable from its environment, and introducing a "discontinuity" in this environment, as in the example of a hill in an altitude area? Or is it preferable to envisage the city from an object perspective? In this case, it concerns an object of the *fiat* and *composite* type, constructed from lower level objects or fields (Figure 1.2). It is then important to recognize the type of relationships and criteria that will be taken into account in the case of a morphological definition to identify where in the built space are located the discontinuities that allow demarcating the city.

a) b)

Figure 1.2. *The morphological city: a peak in the density surface or an aggregation/composition of buildings. For a color version of the figure, see www.iste.co.uk/mathian/spatiotemporal.zip*

In practice, the first morphological delimitations in France have been made on the basis of aerial photos (field source), and the discontinuity of built area is detected through a proximity criterion based on a spacing threshold of 200 m. This criterion implies that the city incorporates, starting from the center, all the built-up area, as long as there is no discontinuity of more than 200 m between two buildings. The value of the threshold (200 m) resulting from institutional recommendations, makes reference to people's space practices. This threshold can obviously be questioned and takes a different meaning depending on the context. Nowadays, the approach can integrate new sources. They may be of the object type, such as the topographic data base of national institutes that include the localization of buildings, or of the field type (or rather of the "categorical coverage" type to use Plewe's terminology [PLE 98]) such as the coverages of land use produced by interpretation of a satellite image. For each new source, the questions of the choice of relationship measurement (distance, adjacency...) and of the value of the threshold associated with the discontinuity concept have to be raised. Thus, the city object is the result of a modeling process, either of composition of lower level objects (buildings), or of extraction of the "built" category of the land use coverage. Figure 1.3 illustrates this approach: the construction of such a *fiat* and *composite* object is made from conceptual reflection (conceptual domain) by operationalizing them through the implementation of a generic process (model domain) mobilizing observable data (empirical domain). Depending on the type of selected source, it will be necessary to assess the capacity of the source to "meet" the model requirements in terms of data. For example, if the choice to build the city-object is focusing on the extraction of built patches [GUE 12], it will be necessary to ensure that the resolution of this source is compatible with the level of resolution associated with discontinuity as it is defined in the model, namely 200 m. If there is a unique representation of the object in the concept domain and in the model domain, this representation is no longer unique in the empirical domain, because it depends on the sources and associated parameters.

If we continue with the example of the city, this time considering a functional point of view, namely that of an urban employment center attracting the labour force of a residential area, then other questions will come up. This design introduces the concepts of "center" and "area" and the fact that the relationship between these entities should be measured by the flow of individuals resident in the *area* and working in the *center*. The model becomes more complex, introducing new objects and new observables, which multiply the possible ambiguities (definition of a center, of an area being attracted) and the number of possible choices when passing to the empirical domain: which indicator to adopt to identify "attracted" entities and what threshold to associate with it? This is more often specified by coming and going between the model domain and empirical domain. The separation between these two domains ensures the possibility to verify the adequacy between the conceptual object and empirical object. However, it is not without ambiguity. For example, in the case of the morphological city, the threshold of 200 m, which relates at the origin to an empirical knowledge, is used in the model domain almost as a standard, given the generality of the use of this threshold in a variety of countries. In the case of the functional city, the indicator most often used to measure the intensity of the attraction is the share of the labour force living in a place and working at the center. The most used threshold for this share in the literature concerning the definition of European urban areas is 10%. This threshold, unlike the former, has no generic character to the extent that it depends on the mesh in which the information on commuters is captured. Therefore, this threshold appears in the form of parameter in the model domain, and its possible values are studied in the empirical domain. The objects constructed following this approach are of the *fiat* type. The specifications of the whole process that leads from story to measurement must be explicit in order to verify its consistency.

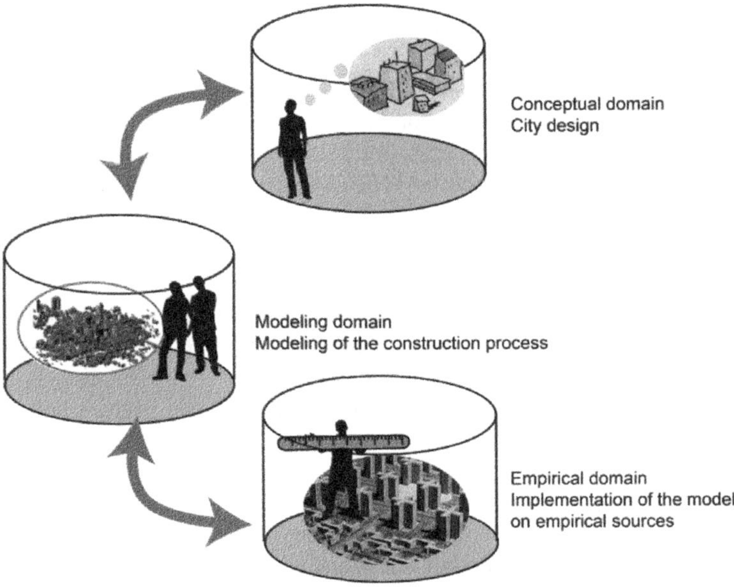

Figure 1.3. *From the conceptual object to its measurement: three steps*

1.2. Locating spatial objects in time

The basic concepts to apprehend the spatial objects of interest being defined, the question is about situating these objects in time. The challenge is to consider them in their full spatio-temporal dimension, by focusing on their evolution as well as on the processes leading to their transformation. In a parallel way to the approach adopted in the first part of this chapter, we will proceed in two stages, first examining the point of view of the philosophers, and then that of the geomaticians, in order to benefit from the additional insight of these two approaches.

1.2.1. *Objects' formalization in time: "endurant" and "perdurant" entities of philosophers*

The dichotomy that the philosophers introduce between the "endurantist" and "perdurantist" conceptions constitutes an enlightening starting point. It consists of two ways of apprehending the world, following the way entities' relationship to time is conceptualized. These two designs are sometimes regarded as opposed in philosophy, giving rise to debates (for example [LEW 88, MAC 02]), but they can also be interpreted as complementary. From a thematic point of view, when there is a need to understand the dynamics of the geographical space, it is indeed the complementarity of these designs that is relevant. In such a spirit, Grenon and Smith [GRE 04] proposed to distinguish between the following two types of entities[9] (Figure 1.4):

– The entities called *snap* (also called "continuant" or "endurant" by philosophers) are entities that have the capacity to persist in time (hence, the term "endure") and that we can observe at any moment of their existence[10]. Therefore, they exist in their entirety at each moment, maintaining their identity. They have an extent in space (in other words, they have "spatial parts") and make up a whole, observable at each moment. However, they do not have "temporal parts" [HAW 04]. Individuals, trees, rivers, cities and administrative units are entities of this type. An individual being does not have "temporal parts", and if he/she evolves over time, it retains his/her identity (Wilson the child and Wilson the pensioner refer to a single person) [LIV 09]. The same happens for a city: Lyon retains its identity even if the city of the middle ages has little to do with the city of today. It exists and can be observed, characterized, at any date of its existence, from Roman antiquity to nowadays. Properties (such as age, surface area and density) are also considered as "endurant" entities. These entities depend on the objects they are associated with (the age of an individual, the length of a river, the number of inhabitants of an administrative unit and the density of a city) and do not have an existence outside of them in the case of an object approach. In a field approach, a cluster of points verifying the same property, defines a spatial extent associated with this property (for example, the set of points having the same altitude that allows identifying the equilibrium line of glacier and following the evolution of a glacier in time).

9 Within the meaning of the definition given in section 1.1.1, i.e. the objects, properties, relationships, events and processes identified to describe and represent an area of interest.

10 The categorization introduced in the first part about *bona fide* versus *fiat* entities, also makes sense in this context. When reference is made to the ability of a *bona fide* entity to persist in time, its existence refers to its physical presence whereas for a *fiat* object (i.e. a built object), its existence concerns the time of the convention (for example, Yugoslavia) or of the conception.

– The entities called "*span*[11]" (also named "occurrent" or "perdurant") refer for their part to events and processes. It concerns entities that have an extent in time. In fact, processes do exist only because of the succession in time of their temporal parts [GRE 04, LIV 09]: they are characterized by a beginning and an end, and a succession of time intervals, each corresponding to a change. If such a *perdurant* entity is observed at a given time t, its entirety cannot be accessed. As a matter of fact, what would aging, growth or acceleration in a time snapshot mean? An entity, such as a process, *is built in* time, it *rolls out* in some way, "it IS its history" [LEN 09]. However, the events correspond to a phenomenon that can be regarded as punctual. While event and process are often distinguished by differences in duration, Galton and Mizoguchi [GAL 09] instead insist on a difference of temporal delimitation, the process spreading over time (outflow of water in a river and spreading of the built area) while the event is dated (for example, fusion of two communes or avalanche). Let us notice that the difference between an event and a process is a matter of perspective and choice in the granularity of the observation. Therefore, the avalanche is an event at the spatial and temporal scale of the lives of the inhabitants, but a process at the scale of the monitoring of the snow along the slope during the seconds or minutes that it lasts. Finally, in some cases, the event marks the beginning or end of a process (for example, starting a new construction).

From the point of view of the domain expert who is interested in the evolution of a spatial phenomenon, who is in other words adopting a dynamic perspective, the two approaches are complementary, each offering a vision on the phenomenon of interest. An individual is a "snap" entity, his/her life trajectory is a "span" entity: this trajectory is composed of different phases of youth, maturity and old age of the individual. A city is also a "snap" entity, it exists at each step of time (of its existence) maintaining its identity. Its "growth" process, as it can be measured by the evolution of its population, of its economic activity or of the extent of its spatial insertion, for example, is a "span" entity. For mobile objects, the "span" conception of certain entities is required from the outset: such is the case of a crossing (for example, maritime), a ride or a pedestrian walking, whereas the objects associated with them, the vessel that crosses the sea and the individual who walks on the street, are "snap" entities. A double temporality must sometimes be envisaged, for example, for a vessel, the one associated with its aging process and the other corresponding to the crossing from one port to another.

11 In English, "SNAP" has the meaning of photo, snapshot, whereas the term "SPAN" refers to the duration, to the scope. This play on words puts the emphasis on two complementary sides of temporality [LIV 10].

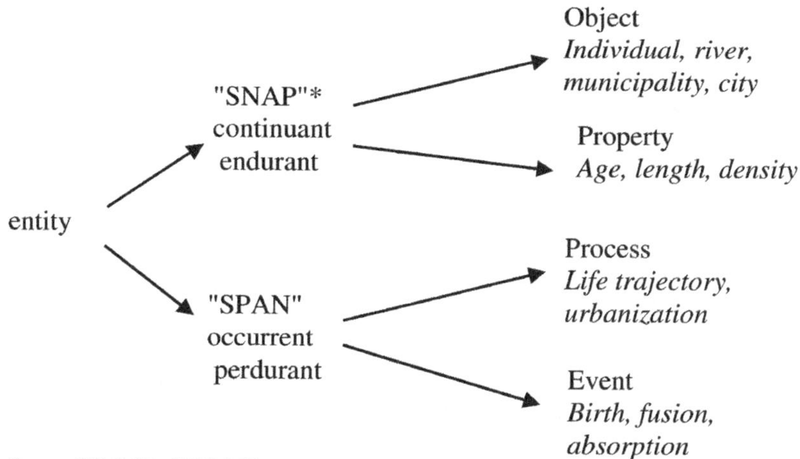

According to [SMI 94, GRE 04]
*: the totality of the terms used in the literature is included in this diagram as synonyms, but it should be noticed that for some philosophers there are nuances.

Figure 1.4. *Categories of entities from the ontological point of view (in italics, some examples related to a variety of objects (human beings and spatial units)*

As highlighted by Bittner *et al.* [BIT 04] in an article proposing to develop an ontological theory encompassing both the spatio-temporal process and the *endurant* entities that are involved in it, the two entities are formally linked: a slice of the *perdurant* entity corresponds to a state (in terms of existence and property) of the endurant entity associated. Figure 1.5 illustrates the two points of view to characterize the phenomenon of urban sprawl for a city in space and time. Galton and Mizoguchi [GAL 09] advocate a related position, stressing the mutual interdependence between matter and objects, on the one side, and events and processes, on the other side, (that is to say, roughly, the two sets of entities represented in Figure 1.4). Indeed, aging would not make sense without an object undergoing this process. Table 1.2 proposes for a few thematic examples, couples of *endurant/perdurant* entities that make sense together. On the one hand, the objects with their properties, and on the other hand, the trajectories that these objects describe in time under the action of processes and events. Vessels, such as endurant entities, are thus associated with the crossings that they perform, which are perdurant entities. Lakes and their saline property, endurant entities, are for their part associated with the process of salinization, perdurant entity that implies an increase in the salinity rate over time.

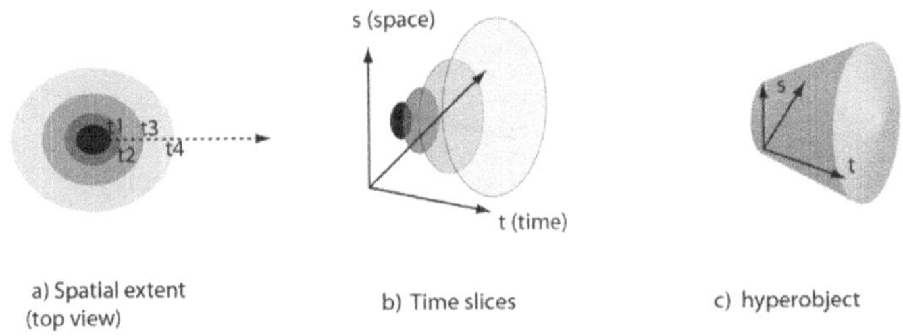

Figure 1.5. *Three representations of the urban sprawl of a city*

Terminology / Entity types	Endurant Continuant "snap"	Perdurant Occurent "span"
Ontological point of view	Objects (fixed or mobile) Properties	Trajectories Processes and events
Thematic point of view	– Individuals (human beings), ships, herds; – Cities; – Countries; – Forests, lakes; – Roads.	– Rides, crossings, journeys; – Urbanization, congestion, growth; – Creation, recomposition (former Yugoslavia), disappearance; – Bushland expansion, salinization; – Traffic.

Table 1.2. *Entities' various relationships with time: ontological and thematic points of view*

Galton [GAL 04] relates these *endurant* and *perdurant* entity conceptualizations to the way of conceptualizing time, space and space-time. Therefore, he extends to time the dichotomy made between field and object approaches mobilized above for space. Thus, he presents a parallel between the spatial concept of field and the temporal concept of duration, on the one side, and between the spatial concept of object and the temporal concept of event, on the other side. This approach results in two conceptions to take into account simultaneously space and time. The first can be qualified of space-time and corresponds to a conception often noted "3D + 1": the objects are considered with their coordinates (x, y and z), referring to their position in space, and then time is added. In this conception, objects can be followed in time as well as in their transformations. This approach corresponds to the case where time-space is designed as existing *a priori*. Objects, events and processes are then positioned inside. It is an *endurantist* vision. The second refers to a conception denoted "4D" in which objects, events and processes define together hyperobjects. This conception is compatible with a *perdurantist* vision. The consequence is that there is no *variation* in time or space, it is integrated in the hyperobject (see Figure 1.5).

1.2.2. *From change to objects' life*

Research in the area of integration and formalization of time in information systems are not new, and the 1990s mark the beginning of specific formalisms with respect to the introduction of time into geographical information systems. If these last 20 years have seen a good number of conceptual and operational formalization developments, a true gap was observed between these advances in the field of research and their integrations into geographical information systems [OSU 05]. These systems have been strongly modeled by the cartographic culture consisting of presenting territories' states, the most up-to-date, and they still leave today to the responsibility of users the care to generate evolution and change. This is done most of the time by multiplying static layers for each date.

In recent years, these attempts were enhanced by reflections on the ontological categories previously presented. Without making a detailed review of all the formalizations developed, nor of the computer developments, for which readers are referred to specialized analyzes [ABR 99, YUA 01, WOR 05, PLU 11], we propose here to walk through the different stages of these developments. Indeed, they raise interesting questions, some of them still remaining unanswered.

The model corresponding to cartographic use, consisting of giving an updated status of a map, is the snapshot model. Change can then only be identified by comparing two successive states of the map. That is often still the used model. However, the first foundations of the issues associated with the introduction of time in order to identify change have been posed since 1988 by Langran and Chrisman [LAN 88]. They already pointed to the fact that a representation of the dynamics by simple superposition of dated layers ("stamped") represents only an accumulation of states where time only exists through the interpretation of the user. They proposed a "Space-Time composite" model based on the differentiation between the status of the map and those of the objects (versions). That way they introduced a differentiation between the changes at the object level (mutation) and the result of these changes at the map level (event) (Figure 1.6). The data model that they propose is change-based in the case of a field-oriented approach. If the field's elements are pixels, each pixel is described by the sequence of its states. In the case of a non-regular partition of space for example, polygons of land use, these are decomposed into elementary polygons according to a principle of uniqueness of history. The elements constructed that way have stable geometries in time. Figure 1.7 illustrates these elements in the case of a city's morphological extension. The different states of the city in time will be described by a series of polygons of built-area: the historical center by t0, the center and the first periphery by t1 and so on for the different urbanized peripheries. However, the elements formed (called "amendment vectors") have no meaning *a priori*: they are simply associated with a story.

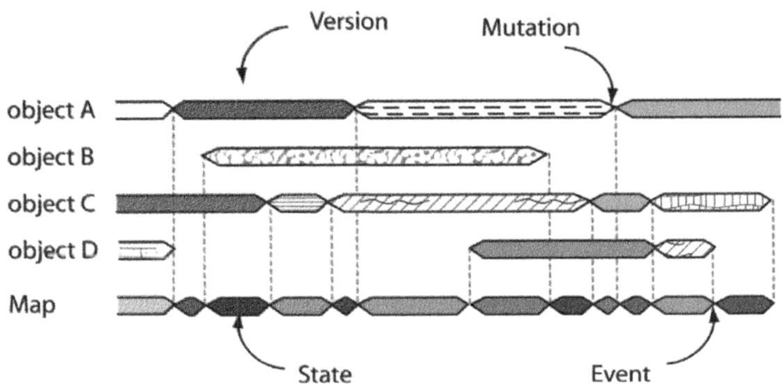

Figure 1.6. *The relationships between map states and object versions (according to [LAN 88])*

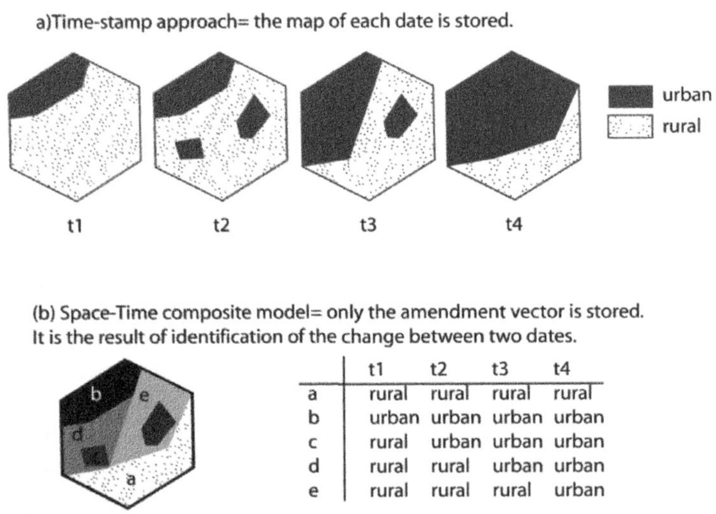

	t1	t2	t3	t4
a	rural	rural	rural	rural
b	urban	urban	urban	urban
c	rural	urban	urban	urban
d	rural	rural	urban	urban
e	rural	rural	rural	urban

Figure 1.7. *The space-time composite data model (according to [LAN 88])*

In this model, the focus is, therefore, above all spatial: changes are only envisaged relatively to the geometry (shape), and the temporal dimension is regarded as a dimension that allows identifying/controlling the definition of the entities. Furthermore, the model is not adapted to an "object" approach.

From this pioneering attempt, a number of works have succeeded, some constituting conceptual formalizations, others leading to operational developments. We retrace the major steps because they constitute a set of interesting elements of reflection. Each development indeed, allowed experimenting with a point of view and going further into the conceptual and computer representation of a spatio-temporal phenomenon. The developments have focused simultaneously on the formalization of change and that of the temporal dimension of a phenomenon, and this in a formal framework incorporating the duality of "object" and "field" representation [PEU 88]. Thus in 1994, Peuquet proposed to extend the formal dual framework associating the "location-based" and "object-based" to the "time-based" temporal dimension. This framework places the triad framework at the forefront enabling the identification of *what, where and when* [PEU 94]. Furthermore, this framework formalizes the fact that our knowledge about phenomena is built through the prism of three subcognitive systems that operate in different ways, but cannot operate independently of each other. By allowing the positioning of a phenomenon according to the spatial and temporal dimensions, this framework allows in particular introducing the notion of event and representing it relatively to these dimensions. Conceived that way, a system must be able to record all types of changes. This framework has constituted a base for contemporary developments and the following ones. Time is introduced in it with the same status as space. A phenomenon can be conceptualized according to these 3Ds that find a simple correspondence in the empirical domain and guide the structuring of the information according to the three components: theme/time/space (Figure 1.8) [PEU 94, YUA 99, MEN 00].

Other developments have committed themselves to express change, to define primitives relatively to each of the three components. Therefore, change may refer to the existence of the object (appearance/disappearance), to its spatial properties (localization/form) or to its thematic properties (qualitative or quantitative). These last two changes are often linked: the type of culture of a plot can change without modifying its shape, but a change of shape usually causes a change in the quantity produced, for example.

This is, for example, the case of the model proposed by Cheylan [CHE 95]: the object is defined by its identity, its spatial extension and its thematic attributes. The monitoring of its evolution needs, therefore, to integrate the change relative to each of these three aspects: the movement for the spatial part (the movement is here referred to in the sense of change in shape and/or localization), the genealogy for the identity part and life for the thematic part. Figure 1.9 illustrates these three aspects of change according to the interrelations that exist or not between the entities: – interrelations in space (case of the partition where an entity's change of geometry implies that of one or several others); – interrelations in time (case of the genealogy where a given entity is depending on the entities it is generated from).

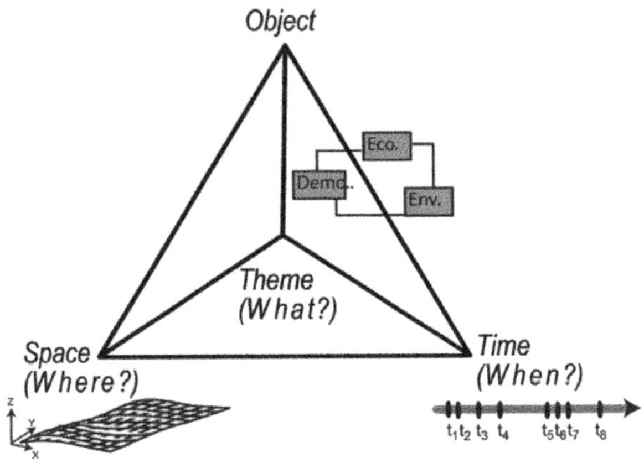

Figure 1.8. *The triad framework*

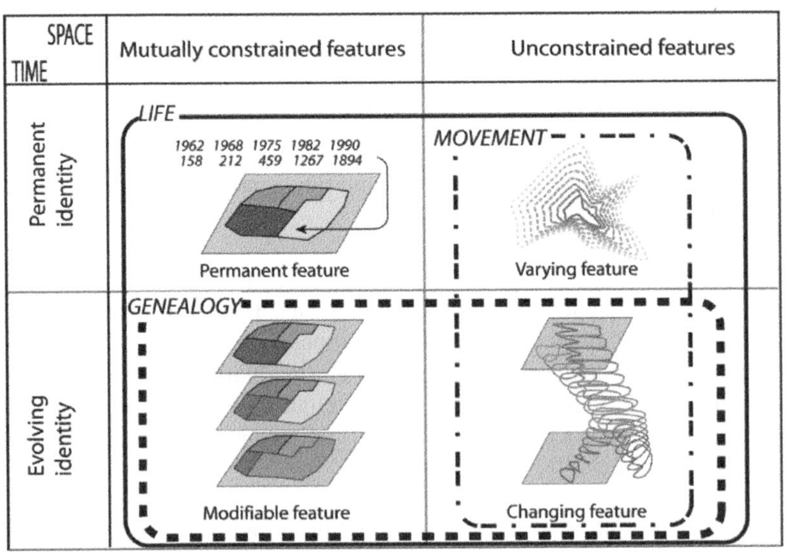

Figure 1.9. *Types of change and categories of objects (from [CHE 93, CHE 00]). For a color version of the figure, see www.iste.co.uk/mathian/spatiotemporal.zip*

Therefore, among the various questions about the representation of change in data models, that of taking into account the interrelations between objects in the description of change appears to be paramount. To the triad "what?, where? and when?" is added the opportunity to question the "how?" [CLA 95]. Claramunt and Thériault propose a typology of what they call "the spatio-temporal processes" into three groups (Figure 1.10). To the first group that concerns the evolution of an object, they add two groups identifying the changes related to the relationships between objects: – a group concerning the relationships that are associated with genealogy and spatial diffusion; and – a group identifying the changes affecting the structure of a set of objects. The geographical object is here considered from a systemic perspective, any change affecting it is likely to have repercussions on other objects (for example, deformation of a border).

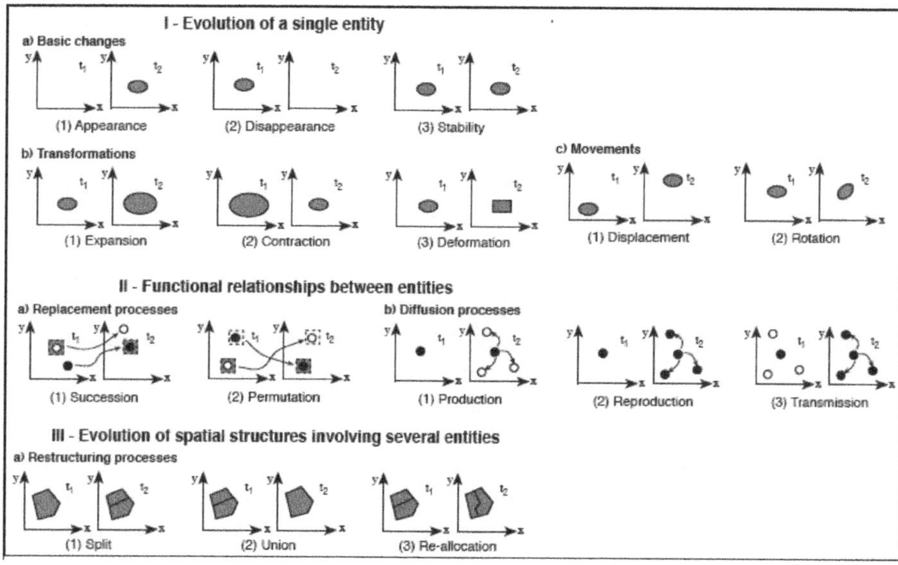

Figure 1.10. *Typology of spatio-temporal processes according to Claramunt and Thériault 1995 [CLA 95]*

This enhancement proposed by Claramunt and Thériault is fundamental to the extent that it clarifies, through the categorization, the relationships that objects maintain with other objects: here, this only concerns relationships of genealogy (successions) and topology, but all types of relationships between objects can be concerned, as well spatial relationships (proximity) as belonging relationships (to an upper level geographical entity) and as functional relationships (influence, exchange...).

From an operational point of view, multiple computing formalizations have been created. Usually developed for a specific question, they have mobilized "object" or "field" formalisms of space. Developed approaches are based either on the identification of objects' change or on events definition and identification of associated spatial and temporal componants [YUA 99, YUA 01, PLU 11]. Consequently, the "Spatio-temporal Object Model" proposed by Worboys [WOR 94] describes the change of an object on the basis of spatio-temporal atoms: the aggregation of these atoms forms the object in 3Ds (two dimension (2D) for space and one dimension (1D) for time). This model is suitable for an object-oriented approach. In parallel, Peuquet and Duan proposed in 1995 a field-oriented model, "Event-based Spatio-Temporal Data Model", which models change through the events associated with sets of localizations (pixels) (ESTDM) [PEU 95]. We can also mention the model developed by Raper and Livingstone [RAP 95], the "Object Oriented geomorphologic model" proposing an object-oriented system based on points integrating geomorphological events. The spatio-temporal data model (STDM) model proposed by Wachowicz [WAC 99] can also be quoted, object-oriented and inspired by the developments of "Time Geography"[12]. It proposes a very original framework of objects' representation: by merging the spatial and temporal dimensions, the objects are represented according to a spatio-temporal trajectory in which states, events and changes succeed to each other.

The totality of these developments leads to approach the concept of identity with precision, and to refer to the objects in their *endurant* characteristic. Some developments have focused more specifically on the question of the object's identity, reaching beyond the single notion of identification and giving it a more philosophical meaning, just like the case of the ship of Theseus[13]. This question essentially concerns the object-oriented approach. When do the alterations on each of its components affect the object itself, namely its identity? The change may affect only the characteristics of the object, or the object itself: the change of culture of a land plot changes its status without changing its identity. What happens if the plot changed ownership? If the single point of view of the plot is considered, then the owner may only be a qualitative attribute of the plot. But from the owner's point of view, ownership will be a determining factor in the definition of the plot's identity, and this change can then be regarded as a creation, after the destruction of the previous one. The notion of identity then becomes central. Several works have been

12 The "time-geography" paradigm, introduced by Hägerstrand in the 1970s in Sweden, will be illustrated in Chapter 2, section 2.4.3 and discussed again in Chapter 4.

13 The metaphor of Theseus's ship has been used by philosophers since the antiquity for questioning the identity and change from an ontological point of view [LEN 09]. The story is told by Plutarch: the ship of Theseus would have remained docked in Athens, the rotted planks being changed gradually. Once all the planks have been replaced, is it still the same boat? And what if the worn planks had been kept aside and then used to reconstruct the boat, "which would be the true ship of Theseus?" [LEN 09].

developed around these issues of identity [REN 96, HOR 98]. Renolen proposed formalizing these changes with graphs illustrating the genealogy of objects ("history graph") and introduced more complex changes as the division or union of objects. Are the 13 states composing "European Union" the same object as the European Union composed of 27 states? What happens to Germany's identity when it was divided into two states in 1949? And at the reunification in 1990, do we find the same object as before 1949? Or still, how can we handle the identity of a forest fire that moves, changes shape and divides itself into several centers? These questions go beyond mere computer management of identity and refer to the intrinsic nature of the object. Hornsby and Egenhofer [HOR 98] introduced the concept of identity states (existing, that never have existed and having already existed) and proposed a visual language to represent the states and the transitions between these different states. The objects may appear (birth or sometimes reincarnation) or disappear; it is a question of grasping their "presence". Germany could be modeled as a "having already existed" identity state from t = 1990. The changing ownership plot could be modeled as a new object created from another destroyed one.

The following positions raise once more the question of what it means to "integrate time" in geographical information systems. While the first attempts of developing "temporal GIS" drew upon the developments made in the area of databases [ABR 99], some authors agree, however, to say that the representation and reasoning about geographical dynamics require more than the simple introduction of time through the notion of change. Indeed, such approach fits in an *endurantist* perspective (vision called 3D+1, 3D for space and 1 for time) while it should be necessary to reflect on the formalization of *perdurant* objects (4D) [YUA 01, REN 00, GAL 04, WOR 05]. Galton demonstrated that such a perspective would allow reconciling "object" and "field" approaches from the spatial point of view, as well as from the temporal point of view. The object approach enters into time through events, whereas the fields approach enters through processes. The challenge is then to make several points of view coexist: a cyclone could, according to the observer, be represented in the form of an event or of a process, in the form of a spatial object or defined by the values of a space field.

The first developments associated with time have introduced the need to work on change, and then on the events and processes, in a "data management" kind of vision based on the background history record and the reconstitution of evolution. These different experiences rooted in the formal development of databases lead then to the need to return to a conception of events and processes envisaged under the thematic perspective, by linking them to the identification of the causes of space's transformations.

1.3. Conclusion

The insights provided by the developments covered in this chapter are fundamental in the way that they guide a conceptual reflection upon the objects, their properties and their relationships and the way that this reflection should be connected to the question being asked. These concepts will be running through the following chapters. The computer environments, in which the various formalisms that have been discussed here are then implemented, do not have neutral roles. The same is true for the formalisms associated with the methods that will then be used to describe and analyze change. The conceptualization of objects in a dynamic context constitutes, in our opinion, the necessary basis to allow using the rich potential of methods and technical environments, without distorting the essence of objects as well as questionings that are dealt with.

Many categories have been introduced in this first chapter. The first concerned *what* is changing, the following *how* change occurs. Some categories are conceptual (*bona fide*/*fiat* objects, *endurant*/*perdurant* entities), others are methodological (field/object approaches, simple/composite entities). Facing a thematic question, reference to these categories facilitates the identification of the entities at stake and the evaluation of the different possible manners to represent them. Reflecting on categories helps to evaluate the consequences of different points of view and different methodological choices in front of a given question. Therefore, such reflection at an early stage of the research is useful to ensure consistency between thematic questioning, concepts and data.

2

From Empirical Questioning to Spatio-temporal Modeling

The common point of the empirical questions addressed in this book is their double anchoring in time and space. The themes addressed in the examples are by contrast varied. They have been chosen on the basis of the interest caused by the construction of the objects relevant for the question posed, and the properties, relationships and processes at stake to answer this question. Another criterion was to show the variety of modeling forms associated with the questioning. From a thematic point of view, this will lead us to examine the evolution of the differentiations between places, to reflect on the representation of mobile objects' paths, to model the dynamics of cities or to represent the temporal frequency of change in an archaeological site. Regardless of the questioning, the aim is to produce new knowledge from observation, processing and analysis of empirical observations related to the topic of study. Chapter 1 was concerned about the conceptualization of objects of interest, with the objective of a coherent construction of these objects during time. This chapter is dedicated to the different stages of scientific or operational work that follows this phase of conceptual construction. The first stage relates to the construction of *empirical entities* corresponding to the *conceptual entities* previously designed. This requires selecting the relevant information and conceiving the adapted data models to achieve the first queries on the data [PEU 02, AND 05]. It then requires choosing and developing the methods that will enable extracting knowledge relative to the targeted subject and answering the questions posed. The objective of this chapter is to explain the nature of the models that intervene in these different stages and the shape of the round-trips that are operated in practice along with this approach. The different steps between the conception of the objects to their analysis will be considered.

Geographers have been greatly inspired by the field of statistics, especially since the 1960s, for the "analysis" part. In particular, spatial analysis has been driven by methodological developments based on descriptive statistics or statistical models. The practices have since greatly diversified; first with the evolution of technologies and computing methods, and then with the explosion of monitoring capacities and the multiplication of data that has accompanied it. Geography has not been immune to this development. It has found itself questioned by the whole acquisition and processing device of all new geolocalized data. This involves the images and data from new sensors, such as Global Positioning System (GPS): in a few years "we have moved away from the era of the localization of information to that of the localized information".[1] All these developments have given rise to new needs and new practices. These, in turn, have given birth to "schools", and also to a mixture between the developments carried out within different disciplines (spatial data mining, geographical knowledge discovery, geovisualization…).

The question here is not to go further into the details of the different approaches. If there are well-identified schools, upon which the development of environments for specific analysis and modeling is based, the practices can then be "hybrid". We propose in this chapter to situate our approach in relation to this diversity of practices. In the first step, we make explicit various challenges associated with the chain that goes from thematic questioning to the production of knowledge. The explanation of these challenges will highlight the different meanings that can take the word *model*. We will stress the misunderstandings that may emerge in the collective work, especially when geographers, archaeologists, geomaticians and computer scientists make use of the same term without specifying it even though they mean different implicit representations. We finally apply in a systematic manner the various concepts introduced on rather diversified examples of empirical issues.

2.1. From the conception of entities to their analysis of responding to thematic issues

In this book, we will give preference to a rather more hypothetico-deductive approach than an inductive one. The question, formulated in the early stages, will direct the different stages of the formalization and modeling and first of all the choice and definition of the entities considered as relevant (choice discussed in detail in Chapter 1). Each of the following stages corresponds to a "challenge", generic in the sense that it can be found fairly systematically in all the applications, even under varied forms. These challenges fall into four categories:

---

1 Thierry Joliveau at the PUCA seminar on the "dynamic representations of the territories' temporalities", Paris, 5 February, 2014.

– Challenge 1: *to build the empirical objects* corresponding to the conceptual objects identified as relevant relative to the thematic questioning: it means comparing the objects (as well as their properties and the associated relationships) that have been identified from a conceptual point of view (see Chapter 1) with the information that is observable from an empirical point of view. The objective is, therefore, to build the empirical entities that would be the most faithful possible to the conceptual representation that was made. We will focus on the entities of spatio-temporal nature;

– Challenge 2: *to represent and explore the changes and movements*[2] performed at the level of these objects, in particular the changes in their properties and in the associated relationships: the purpose is to show and make these changes and/or movements explicit and to build different representations (calculated and/or visualized) by querying the database. This challenge is related to the domain of database queries, visualization and interactive exploration;

– Challenge 3: *to identify and analyze the evolution of relationships* between the objects' properties (or in other words, of the correlations between the attributes or the indicators characterizing the objects) or of relationships between the objects. It is a matter of building new indicators that allow testing assumptions about the organization of the phenomenon being studied (in terms of spatial differentiations, of exchanges between places, etc.). This challenge refers to statistical analysis and spatial analysis methods;

– Challenge 4: *to identify the processes underlying change:* the idea is to formalize models that bring into play the factors considered as driving forces in change and to implement them in a computing environment. It allows simulating the emergence and the dynamics of the phenomenon being studied and testing different scenarios. This challenge involves computer simulation.

Each of these four challenges implies modeling. But the models associated are in general of a different nature, data models, statistical models, spatial analysis models and simulation models. Some of these models will have a more descriptive objective (data models and statistical models), other more explanatory (statistical models and simulation models). In some cases, these four challenges will succeed one another in a linear manner following the stages of a study or research. In other cases, the approach is based on round-trips. The same model can even be used at different stages, with different objectives. A classic spatial analysis model, such as the gravity model, can for instance be used to verify that the intensity of exchanges between

2 It should be noted that movement may be regarded as a particular type of change, that of the change of location of a mobile object. Movements and changes are in general treated differently in the literature, because they raise different questions in terms of time and space (for example, the same methods are not mobilized in order to study change in a territory or the mobility of a person), and we choose to respect this dichotomy.

places is proportional to the masses of places and inversely proportional to the distance separating them as stipulated in the theory (challenge 3). It can also be mobilized as a method for estimating the value of the exchanges during the stage of data construction (challenge 1). Still, it can operate as a filter that allows highlighting pairs of specific places that exchange more or less than the model would let us suppose (challenge 2). Finally, it can be introduced in a simulation model to account for the dynamics of the exchanges between places (challenge 4).

In the remainder of this chapter, these four challenges are examined in more detail, with an anchor in the operationalization. Then, they will be expounded in a series of examples, thematically very different but presented following the same framework, defined by the succession of these four challenges. The purpose of this part is to explain these various challenges and to illustrate their combinations. In particular, the links of dependency that they maintain will be shown. The choices made to overcome a given challenge have, as a matter of fact, implications on the processing that follows as well as on the type of models referred to at each stage.

2.1.1. *Building the spatio-temporal objects from the empirical observations (challenge 1)*

This challenge relies on the conceptual formalization of spatio-temporal objects considered in Chapter 1. That stage has given rise to a first model, conceptually based. Its operationalization gives rise to a physical data model, suitable for a computing environment in which information will be stored and processed. The various categories presented in Chapter 1 will be used as a formal framework for the presentation of examples. It may appear complex to systematically force the different grids and proposed categorizations to cohabit (object/field, bona fide/fiat, simple/composite and endurant/perdurant). The benefit resides in that the questions that this type of specification leads to facilitate a coherent construction of data. This construction evolves through round-trips between the initial thematic issue, the different levels of conceptual questions and the existing empirical "bricks"[3]. Thus, the researcher/practitioner is lead to question himself/herself concretely: what conception of space is the most suitable to its study? Is there a need to design objects, or is it more pertinent to apprehend space as a continuous field? For example, in the study of the glaciers' evolution, should a "glacier" object be defined and its delimitation characterized (fiat object) in order to follow its evolution? Or is it preferable to adopt a field-based approach by interpolating a glacier mass index at

3 We will denote as "brick" the raw, elementary and observable material. The empirical objects will be constructed from these bricks. In some cases, there will be a bijection between the objects and bricks (for example, the cadastral land parcels), in other cases the objects will be a composition of bricks (for example, a parcel's land occupation will require the aggregation of the polygons of land use, which will be called elementary bricks).

any point in space, which will allow studying the movements and variations of the glacier?

These questions must be raised in a dynamic context: the main idea is to adopt the point of view for which the change that we are interested in makes the most sense. This choice is fundamental because the formalization of the spatio-temporal objects of interest and the empirical information necessary for their construction depend on it.

Depending on the point of view, the objects that ought to be built and analyzed are, as a matter of fact, different. We will illustrate it here, in the context of the study of the infraregional demographic dynamics of a French region, for example, Provence, between 1960 and 2010. According to the question, it may be convenient to represent space in the form of objects, such as the municipalities, which will be then described by their populations in time, or in the form of a continuous field; in such case, we will rely on a chronological series of density grids. The point of view specifies the question and objects to build (Figure 2.1):

1) If the evolution of the spatial organization of the demography is of interest, we could choose to work at the level of the municipality (object) or at the level of a grid cell (field). The grid in this context is a simple tool for the localization of the population. Only the states of the territory are of interest and not the underlying process (such as the mobility trajectories of the inhabitants). Only the representation of the sequence of the population's spatial organization is necessary to follow the evolution (Figure 2.1(a)). This is typically a "snapshot" approach: the aim is, for example, to produce a series of maps presenting the states of the density on each of the dates. A synthetic indicator can also be calculated, characterizing the shape of the spatial organization associated with each date (for example, a coefficient of spatial autocorrelation). In this context, the follow-up of the whole (at the department level) is considered important and not the follow-up of each elementary entity (the municipalities or cells level). The elementary entities are only apprehended in the manner in which they make up the instantaneous state of the whole, and there is no need to reconstruct their history.

2) If the evolution of the territory is of interest, it may also be important to shed light on local trajectories (evolution of the populations of municipalities or grid cells). In this case, it will be necessary to build and represent the "municipal" populations, and each municipality will have as many attributes as census years. The municipalities or the meshes are then the only entities of interest. The states of the municipalities are described by the importance of the population at different dates: the change is apprehended through *endurant* objects whose states are followed up during the period. The evolution of the department will be analyzed through "local

histories", those of the municipalities. The diversity of the municipal populations' trajectories can also be analyzed (Figure 2.1(b)).

3) If the point is to understand the territory's dynamics as a process of exchange between places, the dynamics of the department will be viewed as a process with immigrations and emigrations. The aggregated trajectories of incoming or outgoing inhabitants are then considered rather relevant: how many are gone, and where to? And how many arrived, and where from? The population state at a given moment is no longer envisaged as the result of a simple counting, the objective is to take account of the direction of the migration flows of residents (Figure 2.1(c)). We are then closer to a questioning about the processes. The flow becomes the object of interest, characterized by an origin and a destination: this information existing at the level of the municipality, it becomes an attribute of a "flow" object. These flows are *perdurant* entities.

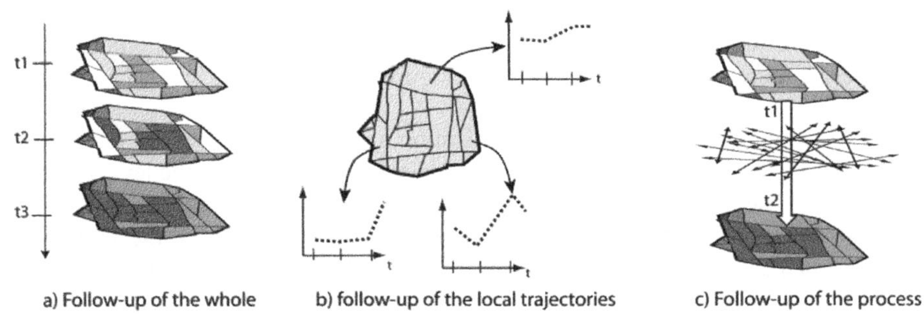

Figure 2.1. *Three points of view on spatial change*

These examples illustrate the practical questions that the conception of empirical objects raises relative to the dimensions of space and time: at what level do we want to analyze change? What objects should be conceived as *endurant* (at what geographical level should time be integrated)? What are the consequences of such choices on data modeling and on the management of identities? Time may be integrated at the object level (the duration of the existence), at the level of the attributes (population at time *t*) or even in the form of events (for example, year of transition from the status of rural municipality to urban municipality).

The result of this challenge is a database where empirical data are organized according to a data model that derives from the conceptual model defined beforehand. This database can be designed in an environment that includes or not a language dedicated to the modeling of spatio-temporal data. Nevertheless, these two aspects, object conceptualization and computing construction, should be considered as two stages. The conceptualization may be envisaged independently of the second. Indeed, it is fruitful to work it all through even if everything is not operationalized.

But, it would be awkward or even sterile to consider the computing stage without the conceptualization stage.

2.1.2. *Representing and exploring change and movement (challenge 2)*

This stage concerns the construction of synthetic "views" of the database. Each of these views relates to a corresponding precise question. This phase is closely linked to the design phase of the database. The questions will have to be made relative to the structure of this database, namely according to a grammar compatible with the formal syntax that represents the data model implemented. In this sense, this stage helps validate the consistency between the data model and questions. As a result, the exploration of data will be facilitated. As has been said in Chapter 1, most data models do not represent change directly, but rather sequences of states or events. The "views" are the result of queries on the database and may involve either a state (where was the Queen Mary on 1st May 2014?) or a change or a movement (what was the itinerary of the Queen Mary in May 2014?). It is necessary that the data model allows answering these questions.

The query is the expression of a question according to a syntax incorporating the objects and their relationships, their position in time and space. It is important to identify in the question the different dimensions mobilized and, in this sense, the decomposition according to the dimensions introduced by Peuquet (space, theme and time), constitutes an extremely useful grid. The questions are being based on these different dimensions, or combination of dimensions: it concerns identifying the dimension targeted by the question (*where? what? and when?*), the other dimensions then defining the reference (that which is fixed) ([PEU 94]). "What" refers to the thematic dimension (an attribute or the very existence of the objects). First, we illustrate simple questions, where only one of the dimensions is queried. Let us consider the case where the objects are municipalities described by their localization (space), their urban or rural status and their population (thematic), over the period 1990–2000 (time). The questions are then decomposed according to the three dimensions (3Ds), according to a syntax[4] inspired by Peuquet [PEU 94] and Andrienko *et al.* [AND 03]:

– Where are the urban municipalities in 2000? (theme +time=> space)

– What were the statuses of these municipalities in 1990? (space+time=> theme)

4 The operator "+" corresponds to the fact that both involved dimensions are linked by "AND". "Δ" indicates a variation that is tested on the following dimension. In this last case, the dimension according to which the variation is calculated, is following, preceded by the sign "/" (for instance "Δ temperature/time" means a variation of temperature over time; while a variation of temperature over space would be written "Δ temperature/space". The sign "=>" symbolizes the result of the query and is followed by the dimension concerned by the result.

– When did a given municipality have at least 2,000 inhabitants? (theme=> time)

The first two queries do not imply the notion of change, time being defined in absolute. The characteristics queried correspond to positions and states, at a given moment. The third query, although having a similar structure, implicitly contains the notion of change. The question targets the temporal dimension ("when"), and finally intends to locate an event in time (the population has reached at least 2,000 inhabitants) to which a process is associated (the evolution of the population). If the event is not formalized as such ("go over the threshold of 2,000 inhabitants" event), then it will have to be apprehended by change; that is by the differential of the targeted dimension:

– When has the population of this municipality exceeded 2,000 inhabitants? (Δtheme/time=>time)

We can express in the same syntax the questions raised to a group of mobile individuals (the inhabitants) described by their localization (space) at a given time (time) and their age (theme). Then, it is the spatial dimension (localization) that evolves in time, and it is, therefore, this dimension that the differential will refer to.

– When was this individual located at this point? (space=>time)

– When has this individual transited from point A to point B? (Δspace/time=>time)

The question may also concern the type of change:

– Where are the municipalities that become urban between 1982 and 2000 (Δtheme/time=>space)

– When have the municipalities of this area become urban? (space+ Δtheme/time=>time)

– Who are the individuals who have transited through this road section? (Δspace/time=>theme).

– When have these young people crossed the river? (theme+Δspace/time=> time)

Finally, sequences of states can be constructed, that will be called history or trajectory. Although change is not explicit in this case, the query will allow its study, directing the selection to the set of states of the dimension concerned by time:

– what is the sequence of statuses of each municipality? How has the population evolved in the municipalities of the 1st periphery of such city?

– what is the sequence of localizations of each individual? What are the sequences of localizations of individuals younger than 20?

If we work on the relationships between objects, for example, the flow of commuters between two municipalities, the sequences can relate to a measurement (1) but also a set of locations (2):

1) What is the evolution of the percentage of the working population residing in municipality A and working in municipality B?

2) What is the evolution of the recruitment area of the employment center C? (i.e. what are the series of municipalities of which part of the residents are working in the employment center C?)

It is obvious that in this exploration of change, the temporal dimension is at the heart of the questioning as it is of the formalization that allows exploring it. If "physical" time is linear [KLE 07], the temporal phenomenon that we observe may appear cyclical (repeating on a regular basis): using a misnomer, we speak of cyclic time. In fact, it corresponds to the way in which we need to formalize the temporal dimension, in order to allow the transition from a linear interpretation of time to a cyclic interpretation. In this case, the granularity of the cycle must be precised. For example, the question "what is the distribution of shops' frequentation during weekdays?" implies that the formalization of time integrates the days of the week. Only in this case, a calculus of accumulation or of average of visits per day, is possible. Other time forms can be envisaged, corresponding once again to other representations of the temporal phenomenon [WOR 05].

These first explorations "view" the data, but they have also another role: they allow us to test the way in which the empirical observations have been implemented and the way in which the objects have been conceptualized and identified in their 3Ds and in their various relationships. We will not exemplify further the exploration of the relationships' dynamics here, whether they concern composition or interaction relationships. The composition relationships (composition of a city in terms of census tracts or composition of a landscape in terms of landscape features) will be illustrated in the examples at the end of this chapter. Analysis of interactions' relationships (material or immaterial exchanges between places) will be processed in the following chapters.

Each of these queries will be able to target either an object or a group of objects, or either a synthesis of a group of objects: we re-encounter the idea of interpretation levels recommended by Bertin [BER 67], the elementary level and top level. When the query targets a group of objects, it will be necessary to use aggregation functions to obtain a synthetic view. These interpretation levels exist for each of the 3Ds. The question may target an object or several objects, a moment or a period, a place or a set of places, an attribute value or one or several intervals. We call the result of a query, a view which, depending on the case may be tabular, graphical or cartographic. If the chosen level is the elementary level on each of the dimensions, the query will produce a view of the raw data (the densities' evolution of a given municipality over all the period's dates). If the general level is chosen for one of the dimensions, it will give rise to a calculation which synthesizes the group of objects of the concerned dimension. If the dimension is thematic, this will be, for example, the evolution of the municipalities' density averages over all the period's dates or the diversity of these evolutions. If the dimension is temporal, this will be the average density of the municipality B over the period.

The form of the presented queries allows identifying states and approaching change by comparison, either visual or by a numeric calculation between two dates. In the first case, this may be the comparison of maps of locations at two dates (question *where*), or the viewing of the cloud of points crossing values from states at t1 and at t2 (question *what*). Secondly, the type of change can also be identified, by evaluating the variation (quantitative or qualitative). This implies having formalized the order structure of time (variation between two successive dates) in a similar manner as the spatial dimension is in geographical information systems (a place and its neighboring places). In this case, we can even express interesting interpretation of the evolution: for example, when the point of interest is the pattern of the urban diffusion, we can seek to identify the municipalities whose rate of population variation between two dates is higher than that of their neighbors.

This stage of exploration, in the sense of sorting, selecting and shaping, is undoubtedly the one that has taken the most importance in recent years. This is related to the increase of volumes of available data (often imperfect and not originating from explicit collection protocols) and to the development of visualization environment. We have seen that these queries can be increasingly more complex and produce statistical summaries of different distributions (spatial and/or statistics). These queries are part of a continuum of processes that range from the simple exploration to exploratory data analysis or statistical analysis (see Chapter 3).

2.1.3. *Analyzing the evolution of statistical and spatial relationships (challenge 3)*

The third challenge requires the implementation of analyses and of statistical models to describe change. At this stage, to describe does not only mean to represent the data as in the previous stage. This is the other end of the continuum of analyses from *showing change* through simple or elaborated queries to *validating* or *invalidating* assumptions about the studied phenomenon. The following example illustrates the difference: for answering the question "what are the average population trajectories of the communes located less than 20 km from employment centers?", one has to design a model that shows urban growth. By contrast, the very similar question "is there a significant difference in demographic growth between the municipalities located less than 20 km and those more than 20 km?" requires the implementation of a statistical analysis.

Particular attention will be given to the statistical approaches to deal with spatial dynamics. A first categorization of the mobilized methods can be proposed:

– Either the concern is about conventional descriptive or inferential statistical methods (regression model and data analysis), used on geographical data, to model, for example, the relationships between the attributes of a set of spatial units. In some cases, the results can be mapped, and the spatial organization is then analyzed a posteriori by the domain expert. Regarding change, either it may be contained in the analyzed variables (series of stamped variables or temporal variations), or analyses are conducted separately for each time step. Here, the results are also evaluated and compared *a posteriori* in order to apprehend change.

– Either the concern is about spatial statistical methods, developed to analyze spatial distributions (for example, nearest neighbor analysis) or spatial configurations (for example, spatial autocorrelation). In the latter case, an indicator (or a hypothesis test) is estimated that characterizes the shape of the spatial organization of the phenomenon that we are interested in at each date. Change can then be interpreted through the indicator being observed during such a summary.

This categorization that is based on the role given to space is obviously very schematic. Indeed, as we will see in Chapter 3, space may also be summarized in the analyzed variables. Another categorization may be proposed in light of the objectives of statistical processing: namely, either modeling in order to infer and predict, or either modeling to describe, to filter or decompose the information.

Furthermore, in statistics, the analysis of time series, namely of observations regularly spaced in time, has been the subject of developments of very emblematic specific methods of the first type mentioned above. The models of time series enable characterizing the distributions in time and making predictions. As an example, the most known should be cited: the class of autoregressive models, based on the regression model and that assumes that the past has an impact on what we are observing in the present, and that the recent past is more influential than the distant past. These models will allow, for example, expressing the Gross national product (GNP) of a country in the current quarter, based on the different values of the quarterly GNP over the last 5 years. Other explanatory variables can also be introduced, measured according to their dynamics: in this example, the series of values of the unemployment rate could be introduced to assess the explanatory power of past values on the current value of the GNP. With these models, the process is not modeled directly by the integration of the different states into the model, but the tendency is modeled and a SPAN representation of the dynamics is being approached. By analogy, a number of spatial models of the same type have been developed, in line of the developments of Cliff *et al.* [CLI 73] on spatial autocorrelation. They incorporate the fact that a value may be influenced by neighboring values, close neighbors having more influence than the remote neighbors. The pairing of these two classes of models has given rise to models allowing the study of the explanatory power of factors evolving in time and space [GRI 10].

Our positioning is a little different: in the approaches that we will describe, statistical models are used to interpret, enlighten and summarize spatio-temporal data, in order to identify change. Here, the idea is to privilege the construction of objects in time and to approach their dynamics in an abductive manner. Indeed, the analyzes are used as successive interpretationfilters, with round-trips between analysis of results, conjecture, and new filter. These approaches combine statistical models and models of spatial analysis, by interpreting systematically the results relatively to the theme. Furthermore, it is useful to identify if space and time are integrated in the analysis, or on the contrary are the support of the results. We insist on the use of the term "approach" that corresponds closely to the intellectual path that the application of a sequence of methods gives rise to, producing complementary insight of the analyzed phenomenon.

Let us consider two examples to illustrate the different roles that space and time can take in the analysis. Clark's model, classical in urban geography, is a static model of spatial analysis, where space is formalized through the variable "distance to the center". It describes the shape of the decrease of the density as a function of the distance to the center[5]. If we apply this model to a same space at several dates, then the result is a sequence of curves (that can also be summarized by the sequence of the two parameters) evolving in time. An indicator can also be calculated: for example, the distance at which the density decreases below a given threshold, which provides information on the sprawling of the city, and permits to follow it up over time. In this example, time acts as a "frame" to interpret the result. Conversely, if a table of the evolution of temperatures over a number of meteorological stations is analyzed and the diversity of the evolution is summarized by the first factor of a factorial multivariate analysis, then time is integrated in the analysis. In addition, space intervenes a posteriori during the interpretation of the spatial organization of the first factor, summarizing the diversity of forms of temperature evolution of the stations. From the point of view of methods, these two examples also illustrate two different ideas for the elementary and general level: in the first example, it is the overall structure that is described at each time step, and the evolution of this structure is being followed. In contrast, the evolution of each station in the second example is being described, and a summary of these dynamics is spatialized.

This decomposition according to the various statuses of time and space, as well as the geographical level at which change is analyzed, allow identifying a series of questions, which will be discussed and illustrated further in full length in Chapter 3:

5 Clark's model is written as $D(x) = a \exp(bx)$, where $D(x)$ is the residential density, x is the distance to the center of the city and a and b are two parameters (a represents the density in the center, and b is the dissuasive effect of the distance).

– to model statistically the evolutions of spatial units and identify the organization of differential evolution in space;

– to model spatial organizations (for example, using spatial autocorrelation) and identify their evolution;

– to analyze and follow-up spatial associations (or in other words, correlations between variables characterizing spatial units) in time.

This challenge may therefore constitute a goal as such or be integrated in a loop. It can for example be used for creation of data and its explorations. Some processings may be integrated in exploration methods. Finally, these approaches can also be used in a preparatory step to other types of modeling.

2.1.4. *Identifying the underlying processes of change: simulation and scenario testing (challenge 4)*

Finally, the last step is the one that identifies and proposes explicit formulations of the processes that lead to certain types of spatial organizations and to their evolution: it concerns the implementation of dynamic models and simulation models. Let us recall the example related to the demographic evolution of a spatial unit. In the most simple case, where questioning focuses, for example, on the evolution of the population of a given municipality K, the purpose is to identify the factors that explain this evolution. Knowing the population at the date being considered as initial, the objective is to build a model that allows explaining the shape of the municipality's trajectory. Various frameworks and formalisms are possible. Here are a few examples:

Aggregated dynamic model

The evolution of the municipal population can be expressed based on a difference equation or differential equation. The logistic model is a classic example:

$dPK/dt = a\ PK\ (1 - PK/CK)$ for the differential equation version, in continuous time

$PK_{t+1} = PK_t + a\ PK_t\ (1 - PK_t/CK)$ for the difference equation version, in discrete time

where PK represents the population, CK is the carrying capacity of the spatial unit considered, i.e. its demographic potential, and *a* is a parameter measuring the overall growth rate of the population. Such an equation translates the hypothesis of a strong growth (of exponential nature) when the size of the population is much below the carrying capacity, and a progressively much slower growth as the population gets closer to this threshold (Figure 2.2).

The formal properties of this model are well known, but it is not simple to determine the shape of the term CK (that is the demographic potential of the spatial unit) from the thematic point of view. In addition, in an evolving system, such a potential is rarely fixed. It depends in particular on the relative place that occupies the municipality in the spatial system in which it is embedded (for example, its position relatively to a larger urban center). In order to take into account the effects of such spatial interactions Allen [ALL 97] develops a system of coupled differential equations.

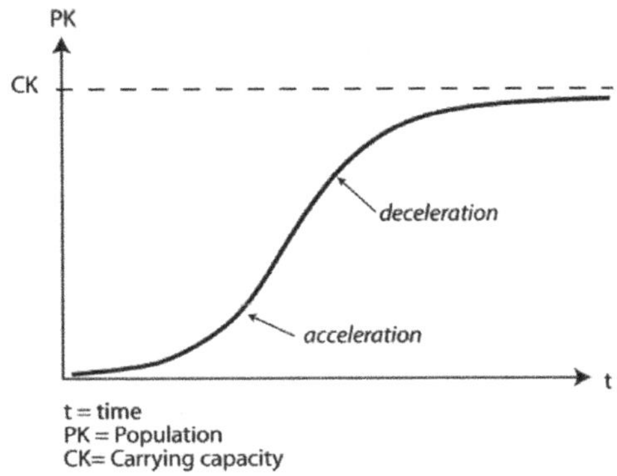

Figure 2.2. *Logistic growth*

Dynamic model of a spatial system

Let us suppose that the respective demographic growth of a set of municipalities making up a system is the point of interest. The underlying hypothesis of such an approach is that the interactions between the municipalities are the driving force of change. The point is then to develop a model that takes into account the growth differentials between the communes. The model, again shaped in the form of global logistics, may in this case be written in the form of a coupled equations system whose general term would be, for example:

$$dP_i/dt = a\, P_i\, (1 - P_i/C_i)\ ,\ i = 1,\ldots.., n$$

$$\text{with } C_i = f\,(E_i,\ A_i/\sum A_j)$$

where P_i represents the population in the municipality i, *a* is the rate of growth (global parameter) and C_i is the carrying capacity of the municipality *i*. This term is expressed as a function of a characteristic E_i of the municipality (for example, its economic potential) and the attractiveness A_i of this municipality (for example, residential potential) relatively to that of other municipalities *j*. This formalization also takes into account the interactions between the dynamics of the different municipalities, particularly through the term A_i that is evolutive.

Microsimulation: an individual-centered model in social science

The two previous cases are based on the hypothesis that a law can be formulated at the level of spatial units apprehended in aggregated form. This assumes either that the individuals' behaviors are homogeneous, or that the individuals' characteristics compensate for each other and that they are seldom determinant relatively to the general tendency. Sherlock Holmes summarizes well this point of view: "While the individual man is an insoluble puzzle, in the aggregate he becomes a mathematical certainty. You can, for example, never fore-tell what any one man will do, but you can say with precision what an average number will be up to. Individuals vary, but percentages remain constant" (Sir A.C. Doyle, quoted by Haag [HAA 89]). Other researchers have an opposite view, stating that it is essential to apprehend logics in change (here fertility, mortality and migration) at the finest level of the individuals. The microsimulation models describe these behaviors with rules operating at the level of individuals, and the evolution of the different municipal populations is then obtained by simple aggregation (see Chapter 4).

Cellular automata (CA) and multiagent systems (MASs)

These methods are also individual-centered from a methodological point of view, but the elementary entities are not necessarily individuals as in the previous case. This will be illustrated in Chapter 4.

The models developed with these formalisms are based on rules. An emblematic example of CAs is the game of life [GAR 70]. The entities are the cells of a grid that can have two states, those of dead or alive. The transition from a state in t to a state in t+1 is defined by the following set of rules based on the neighborhood of the cell: (1) if a dead cell is surrounded by three neighboring living cells, it "is born"; (2) if a living cell is surrounded by two or three living neighbors, it remains alive; (3) if a living cell is surrounded by less than two neighboring cells, it dies by isolation; if it has more than three neighboring living cells it dies by congestion. These simple rules are sufficient to make a great diversity of spatial and temporal configurations emerge. The central mechanism of CAs, which make change in a place (cell) dependent on the neighboring places' (cells')status, meets the basic assumptions of geographers on the functioning of geographical space [TOB 79, COU 85, WHI 93, BAT 05]. Therefore, CA is a dynamic modeling tool privileged in the discipline.

While CAs enable simulating the states and evolutions of space portions, MASs formalize the interactions between autonomous agents located in an environment they perceive and on which they can act [FER 95]. This formalism is particularly well-adapted when faced with a system in which it is assumed that the interactions between the entities considered as elementary (the agents) are the driving force in the dynamics of the system. It follows that these interactions lead to the emergence of a phenomenon of interest that can be observed at a higher level of organization. It involves clearly a *bottom-up* procedure (just as for the CAs). The emerging configuration at the higher level can in turn influence the behavior of agents following a *top-down* logic. The properties and usefulness of these models in geography and archeology are discussed in Chapter 4.

In conclusion, we have presented four issues (or "challenges") (section 2.1) in a linear manner. We will see later, when presenting case studies that each issue "challenge" can have a more or less significant weight depending on the questioning and data. In addition, the sequences are not necessarily linear, and loops can be seen to occur between certain issues (for example, exploration-simulation or conception-analysis...). The whole procedure contributes to highlighting new knowledge about the phenomenon being studied. From an epistemological point of view, the approach is thus abductive, with a multiplication of points of view to better observe the information and test hypotheses on the relationships between different phenomena.

2.2. Challenges and models: the possible misunderstandings

The previous section makes account of the diversity of the types of issues and models that a research about the dynamics of a spatial phenomenon can mobilize. Conceptual models, data models, exploratory models, statistical models, spatial analysis models, explanatory models and simulation models can sequentially follow one another in order to answer to a given problematic but in practice such a succession is rarely the case. According to the question, its field of competence and disciplinary practices, a researcher will in general mobilize a subset of this chain of models. As a matter of fact, depending on the question asked, several sequences of this chain can be drawn. Some questions will give rise to a long construction stage leading essentially to a representation and exploration step of the created database. For others, the stage of the construction of the objects will be short and the analysis or simulation step will be the main challenge. It should also be stressed that this sequence is not always linear. For example, the analysis may lead to the construction of new objects, and simulations generate data that should be explored.

In practice, the term model is often associated with a thematic specification (growth model, hydrographical model or movement model) and less frequently to a specification related to the type of model (descriptive *vs.* explanatory model, or, following another register: data model, statistical model or simulation model). This is without any doubt the reason why misunderstandings can occur in the interdisciplinary collaboration between domain experts, computer scientists and geomaticians. Figure 2.3 provides an example of this case: it illustrates a meeting whose purpose is to analyze "the potential for growth of rural municipalities" in France and the speaker mentioned "the problem of modeling spatial dynamics". Everyone hears the same sentence, but the interpretation is rooted in disciplinary practices: the geographer-modeler raises questions about the form of the process at the origin of the growth differentials between the municipalities; the geomatician, questioning the data model, identifies here a very simple case, the objects being clearly identified and their location and geometry being fixed over time; the computer scientist imagines a structure allowing queries on the growth rates of municipalities. "Modeling spatial dynamics" does, therefore, have a meaning for each one, but these meanings differ, without being in contradiction. They are simply referring to different modeling challenges. The problem is that each of the participants does not always anticipate the different meaning given by the other.

Figure 2.3. *Models and interdisciplinarity*

A second example enables specifying the forms of the dependence of the representations on the disciplinary background. This time, the question is about "modeling the dynamics of a landscape" and two actors are in play, a geographer and a geomatician:

– The geographer thinks about the regularities in the change of land use: what role does the distance to the closest city plays in this change? And the economic and social profiles of this city? How can the role of the actors be formalized (individual *vs.* macroactors)? These questions refer to statistical and simulation models used to explain, relate and test hypotheses about transformations in the landscape.

– The geomatician for his part raises questions about the opportunity of a continuous field or object approach: – what is the most appropriate satellite image in the first case, to identify, for example, the appearances/disappearances of hedges in the Normandy countryside? To observe the retraction of prairies caused by built areas? In the second case, what key objects must be selected? How do we compare the landscape at different dates? The problem is about building a data model to show and explore data by queries, to view the states of the landscape and how they change.

Interdisciplinary collaboration and exchange allow respective enhancement of the points of view: semantic enhancement for the formalization and model specialists, theoretical enhancement for the domain experts who must explain and abstract. These enhancements are only possible if a questions and answers game is developed through systematic interactions. Each one must agree with the needs and goals of such a modeling. As we have already seen, a data model allows recording the versions of an object, but it will be different depending on whether its aim is to take account of a state or to explore change. Furthermore, if it is about following up dynamics, the models implemented will still be different in nature and will have to integrate processes. It is then necessary to mobilize other kind of information than empirical data. These different aspects are complementary. Some may specialize in one specific area, but in order to implement a consistent chain of processing, it is necessary to be aware of the challenges associated with each of the possible other issues of the chain.

2.3. Application examples

The practices are such in the field of geographic data processing, that none of the approaches is the linear sequence of a series of stages. More usual is the combination of models of different levels (data model, spatial analysis model[6], statistical model, visual representations and simulation model). The categorization by "challenges" that we have proposed above is one way to introduce such practices. In this section, we are presenting four examples and declining for each of them the questions and treatments related to the four challenges. The four examples have been chosen for their complementarity relatively to the various points covered so far:

– change or movement processing;

6 Formalized models to respond to spatial analysis problems and mobilizing notably the concepts of neighborhood, proximity, network and spatial interaction. A number of examples will be discussed in Chapter 3.

– the mobilization of a great diversity of sources, tracking sensors, census, surveys, archaeological dig etc.;

– underlying conceptual options especially according to the different ways of integrating time and space;

– the different categories of change;

– the objectives and the different involvement of the four challenges as they were defined in section 2.1.

In all the examples, the description focuses primarily on the first two challenges, namely the construction of the objects and exploration of their evolutions. The other two challenges are only presented in a concise manner because they are the subject of the two following chapters. The examples do not have the same length, the idea being to address a diversity of cases. Some are longer because they are better suited to illustrate many of the points listed above, while others are present to illustrate a challenge treated in an original manner.

2.3.1. *Cities' dynamics: construction and follow-up of composite objects in time*

Urban growth is a particularly interesting case because it illustrates well all the challenges that may arise in the construction process of a complex object, going back and forth between conceptual and empirical choices. It is typical of objects that are not defined in an unequivocal manner [OPE 81].

The example that is presented here is inspired by several projects that may be brought together[7]: particular emphasis will be placed on the problematics of definition and measurement of the city at the European level. To illustrate these different forms of representation and modeling, we will briefly introduce the other challenges that this problem has led to, namely the exploration and analysis of the dynamic database, hence constructed, and finally its use in the context of simulation models of urban dynamics.

2.3.1.1. *Step 1: to build a set of objects coherent in space and time: a harmonized database of European cities*

This question, raised at a very operational level, presupposes, however, that it is possible to define an ontology of the "European" city. Such a challenge calls for a dialog between the conceptual and empirical levels. As has been said in Chapter 1,

7 This example is drawn from multiple experiments led by Bretagnolle: – ANR Harmonie-cités (http://www.parisgeo.cnrs.fr/spip.php?article38&lang=en) [BRE 11a]; – ESPON DB program – urban data [BRE 13] (http://www.espon.eu/main/MenuToolsandMaps/ESPON2013Database/).

the city is a complex object for which there are several definitions, and whose definition is evolved over time. The set of cities studied is that of the European cities. The objective is to build a coherent database to then analyze the different dynamics of the cities relatively to indicators based on population statistics (employment rate, capacity to attract as a labor market, activity rates etc...).

The issues raised by following up cities as "objects" on a long period of time of two centuries, while integrating the functional changes of these objects, have already been presented in Chapter 1. In this example, the focus is placed on the challenges that the construction of a database about cities at an international level raises; that is to say, the challenges of harmonization are not in time here, but in space. Indeed, each country has its own definition, its own administrative level of statistics collection (census, survey and civil status) and the definition of cities is submitted to evaluations based on national statistics. The city is here considered solely in its morphological definition that is based on the continuity of the built area. The first stage consists of assessing the spatial envelope associated with this definition, for each city and each date. The second stage consists of evaluating the associated semantic attributes, knowing that they come from different statistical sources that correspond to administrative elementary entities, the local administrative unit level 2 (LAU2)[8]. These various statistics (resident population, active population, employed population, etc.) will have to be "transferred" through a model of aggregation or disaggregation at the level of the envelope defined by the continuity of the built area. The third step will consist of testing the construction method of the entities in time, and assessing the consistency of the *endurant objects* constructed that way.

If the example of the morphological city is finally a classical example from a geomatics point of view, the fact of raising it in an international context renews this question. For the construction of international databases, two types of approaches can be mobilized: a *bottom-up* approach that consists of building an international database from the national databases, by integration of the different national specifications; a *top-down* method that consists of building an international database from a common source existing at the European level and in deriving the same specifications (Figure 2.4). In the *bottom-up* approach, we will firstly rely on national databases (Figure 2.4(a)) that define the agglomeration on the basis of an aggregation of elementary administrative entities, the LAU2, according to a continuity criterion of the built area evaluated in most cases by photointerpretation. The construction of a database at the international level would require aligning the specifications of the national databases and evaluating the equivalences in order to

8 The LAU2 corresponds to the second level of the administrative nomenclature established by Eurostat for the totality of the European countries. This corresponds, for example, to "municipalities" in France, to wards in the United Kingdom,...

check if it is possible to propose a coherent basis [SHE 04]. After expertise, this solution has not been retained in this application, because some very specific national definitions cannot be harmonized with the others. The *top-down* approach has, therefore, been preferred here (Figure 2.4(b)) by mobilizing information existing at the scale of the totality of the countries concerned, the *urban morphological zones* (UMZs)[9] database created from the CORINE land cover database from the European environment agency [GUE 12, BRE 13].

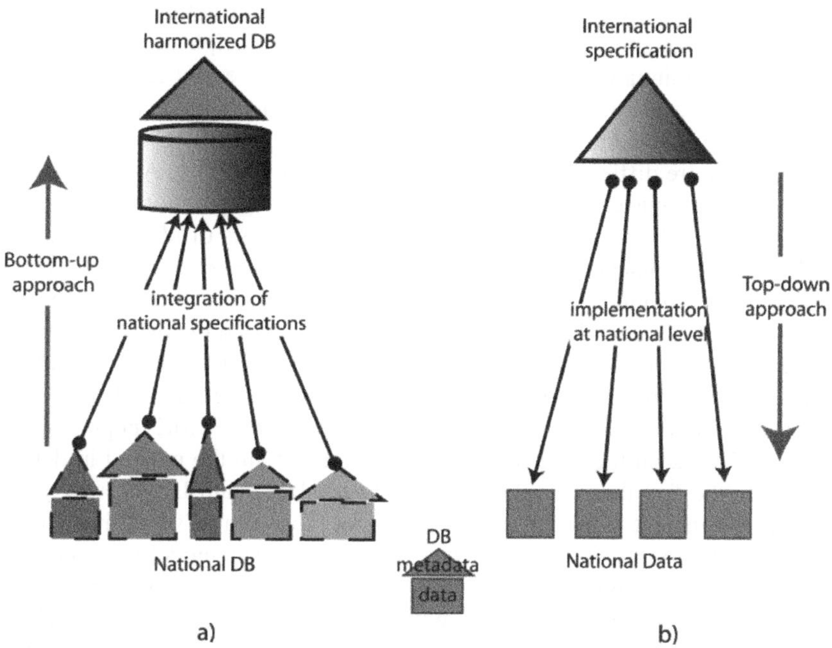

Figure 2.4. *Harmonization of national databases: two approaches (source: [BRE 12])*

Starting from "field" type information (satellite image and photointerpretation) the CORINE land cover database proposes information of the "categorical coverage" type (see Chapter 1, section 1.1.3) for land use. The UMZ database is derived from this information, by extraction and aggregation of the polygons associated with "built" area[10] categories, including also the infrastructure categories. This information is then rasterized (with a resolution level of 100 m). Finally, by a

9 http://www.eea.europa.eu/data-and-maps/data/urban-morphological-zones-2006.

10 http://www.eea.europa.eu/data-and-maps/figures/data/urban-morphological-zones-2000-umz2000-f1v0.

process of dilatation/retraction[11], the areas closer than 200 m are connected between themselves and aggregated. That way the information thus created is compatible with the usual meaning of built area continuity that underlies the definitions of urban areas. Thus, a set of geometric envelopes that constitute zones is obtained, and a map of the urban spots built in this manner can be established.

Identifying the objects

At this stage, the issue of identity is crucially raised. If we are talking in a general way of the city of London, everyone has a representation of it. Faced with a map of the urban spots performed according to the principles described above, a person with a minimum of experience will know how to easily identify London in this set of zones, as soon as she/he knows approximately its localization, and as she/he knows that it is the most extended city of the region [BRE 13]. The question would be yet more difficult for a smaller and less "unique" city. As they have been defined, the areas thus constructed constitute a statistical population whose identifier has no meaning. The second step of this operation consists of identifying the areas according to the nomenclature of cities' names that each country uses. This is a stage that gives semantics to these areas. In order to achieve it, the LAU2 database is used that records for each LAU2, its name, its geometry and the resident population. The operation, which consists of transferring the information to the UMZ database is not unequivocal and needs to define a model to assign the name of a LAU2 to a UMZ on the one hand and a model to calculate the UMZ population from appropriate LAU2 on the other hand. We will not explain the method in detail, but we illustrate the fact that the relationship between UMZ and LAU2 with regard to the name assignment is not a simple bijection. This is not surprising, since the toponymy is the result of historical usages and events. The model relies necessarily on a standard type of process like "the city is built around a center which remains the most populated", but the result will depend on the shape of the city, and therefore on its history (monocentric and polycentric). Figure 2.5 illustrates the diversity of cases corresponding to the combination of these two sources. The model, which was proposed, is based on an index of symmetrical coverage computed on the intersection between the UMZ and LAU2. It is based on what the intersection surface represents in the respective surfaces of the UMZ and each of the concerned LAU2. The UMZ receives the name of the LAU2 entity with which it is most "linked". It receives as population the sum of those of the LAU2 that intersect it at the pro rata of the intersected surfaces.

---

11 This procedure consists of "expanding" all the built-up areas with a thickness of 100 m, so that all the areas closer than 200 m to each other connect and form a single zone. All the areas are then reduced to 100 m in order to recover the initial morphology, not including those areas that will remain connected.

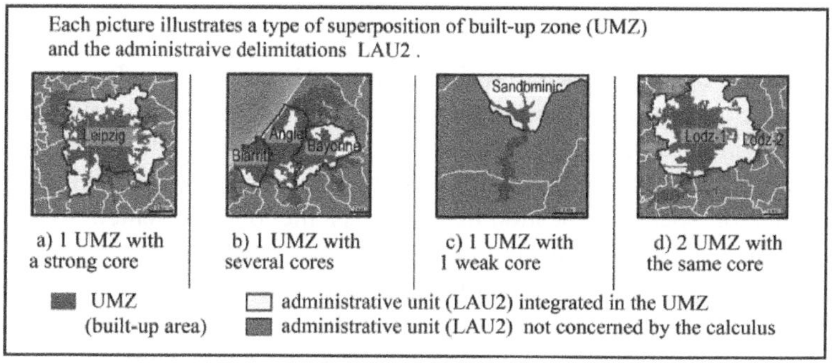

Figure 2.5. *Combination of sources for identifying the cities as objects (source: [BRE 13])*

Time follow-up

The UMZ database is updated regularly by the European Environment Agency. Figure 2.6 shows the versions of this database in 1990, 2000 and 2006 for the urban spots of Poznan and Leipzig. In this approach, which derives from a continuous-field source, the object is only defined a posteriori, after being carved out according to the category of "built-up" land use. Its geometry is determined that way. The object's identification is a procedure in itself and requires an external database. In an "object" oriented design, it is current practice to define the object before, and then to evaluate its geometry and attributes over time. Here, the approach forces to build the geometries and then to identify the cities at each date. This approach does not necessarily guarantee the continuity of the identity. All the changes described in Chapter 1 may be met: appearance, absorption or fusion and division or disappearance. At this level, it is, therefore, necessary to put in place a genealogy system in order to follow the life of the objects. In some contexts of rapid urbanization, it requires some expertise to validate and eventually correct the filiations.

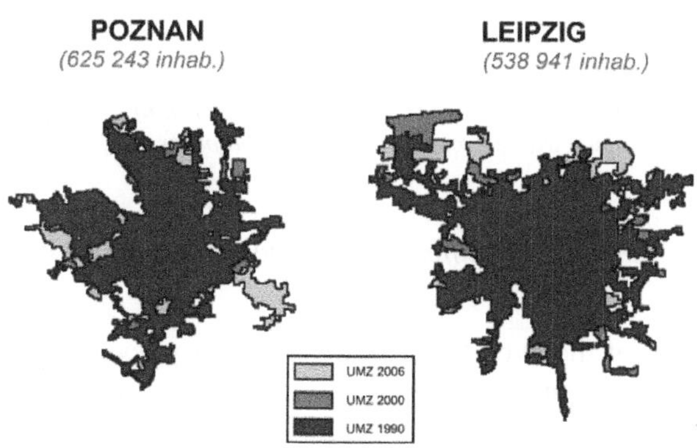

Figure 2.6. *The different versions of the UMZ (source: [BRE 11b])*

2.3.1.2. *Step 2: to explore the dynamics of cities*

The database thus constructed, while ensuring the semantics coherence of objects in time, can be queried in a classic environment of database management. It can also be queried within a geographic information system that allows in addition a visualization of the spatial extensions of the objects. It may also give rise to environments of specific questionings such as the one presented in Figure 2.7 [VAN 10]. Here, it is not an environment for the exploration of spatio-temporal dynamics, but a tool for the visualization of cities' evolution trajectories in an *endurantist* perspective (series of snapshots). Evolution is, therefore, represented under a kinematics form from a simple time interpolation. Its significance lies in the fact that it unveils simultaneously and in an animated way the evolution of a city in its geometric and thematic component (population or surface) at the same time as its inclusion in the local, regional, national or international urban system, according to the need. Possible queries are for example: how is the dynamics of this city in relation to that of the neighboring cities? How is the territory becoming urbanized in this region?

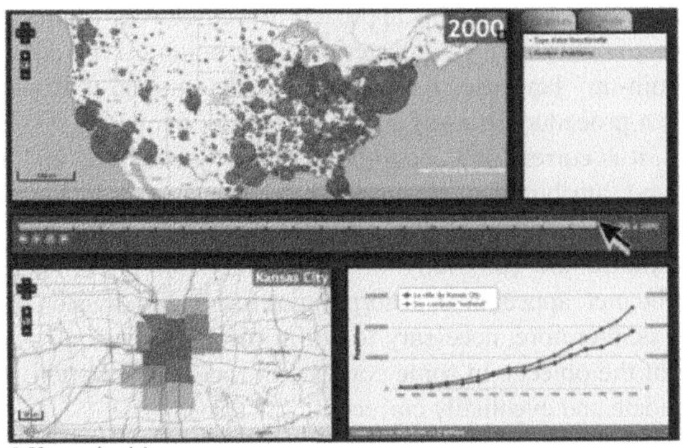

© Harmonie-cités

Figure 2.7. *Animating the cities: exploring the dynamics of cities (source: [VAN 10, BRE 11]). For a color version of this figure, see www.iste.co.uk/mathian/spatiotemporal.zip*

2.3.1.3. *Step 3: to analyze the differences in the evolution of cities: a trajectories' typology*

The exploration of the evolutions at the elementary level of cities then leads to an analysis that presents in a synthetic manner the tendencies of these evolutions. The process, which consists of classifying the population trajectories at the city level, is very enlightening to illustrate the variety of dynamics at stake, as well as any discrepancies and their rhythms. We assume here the point of view that consists of leaving aside the spatial dimension of the evolution at the city level, in order to concentrate on the description of the population's trajectory of each of the cities. The cities are classified in the light of these trajectories, and types will be mapped revealing *a posteriori* the spatial organization of the associated dynamics. These dynamics varies from cities with a very strong growth to cities in decline. The passage from the exploration to the analysis implies a description of the objects in the form of a statistical table; that is to say the cities must "exist" on the whole of the period. Such analysis will be discussed in Chapter 3.

2.3.1.4. *Step 4: to simulate the dynamics of a system of cities*

In parallel to these empirical analyses, a simulation model of urban dynamics has been developed[12] (Eurosim, [SAN 05, MAT 08]), to enable testing the mechanisms underlying the dynamics. Several elements from the three previous steps are mobilized and combined with theoretical elements in order to conceptualize the model in the multiagent system formalism. The database represents the empirical anchor of the model: an extraction of the database at a specific date constitutes the initial situation of the model; the whole of the database will then be used to calibrate the model. At the scale of the whole system of cities, for example, the comparison of observed and simulated dynamics will allow evaluating the sets of coherent parameters. This procedure will then be adapted to refine this set of parameters, to finer geographical levels on the one hand, for the cities at the top of the urban hierarchy on the other hand.

This example makes up without any doubt the most complete and emblematic example of the different challenges of the treatment of spatio-temporal dynamics. If each stage corresponds practically to a project, the fact remains that the whole is paired and that each stage learns from the next. At this level, we could even restart from the exploration and analysis challenges, like a spiral, since the simulation models generate simulated data that the modeler must explore to assess its plausibility.

2.3.2. *Distribution of urban functions in the intra-urban space: construction of spatio-temporal functional objects*

The second example is derived from a study on urban archeology proposed by Lefebvre *et al.* [LEF 12]. Its relevant aspect is that it takes a position in the urban field as the previous one, but at the intra-urban level this time, the city considered in its entirety constituting the system. The study brings a different light to the evolution of the city relatively to the classic comparative approach that consists of superimposing a city's different boundaries, given *a priori* on the basis of historical

12 http://www.simpop.parisgeo.cnrs.fr/applications/eurosim.

knowledge. The authors propose a method placing time and functional aspects at the heart of the conception of the atoms making up the city.

2.3.2.1. *Step 1: to build a coherent set of functional objects in space and time*

The model underlying the construction of the city is the OH_FET model proposed by Rodier and Saligny [ROD 10]. Recalling Peuquet's decomposition (*what, when and where*), the basic object of the database, called a *historical object*, is defined as the largest stable element with regard to the 3Ds: functional, temporal and spatial. For example, a church will be a *historical object* if and only if it does not change either form, or function in the whole of its "existence". Otherwise, it will be decomposed into as many elemental objects that verify these stability conditions on all 3Ds. These historical objects are then projected on each of the three dimensions and allow slicing each of the dimensions according to this projection, defining what the authors call "spatial entities", "temporal entities" and "functional entities" (Figure 2.8). This decomposition approximates the model proposed by Worboys [WOR 94].

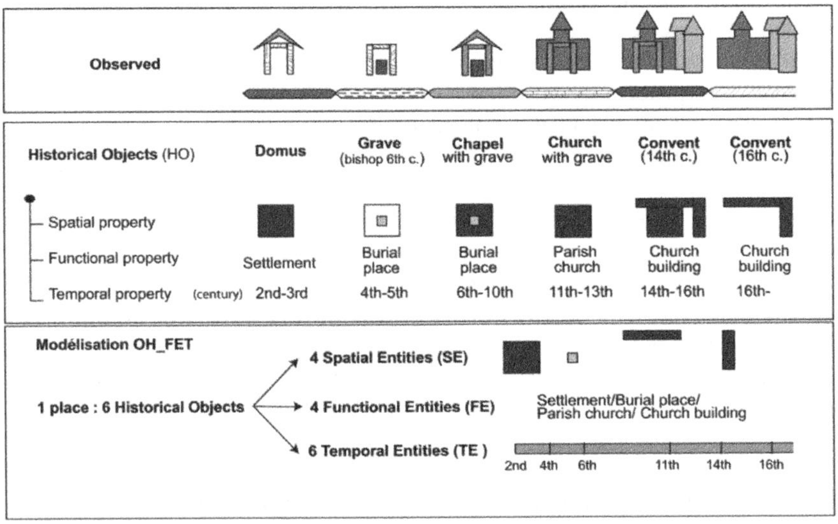

Figure 2.8. *From historical objects to spatial, temporal and functional reference entities (according to [LEF 12])*

2.3.2.2. *Step 2: to explore the temporalities*

The interesting aspect of the *historical objects* (HO) decomposition is to be able to represent the usage intensity of spatial portions and to show the profound transformations that operate on fine spatial scales.

The authors divide this exploration into three scales: the city, the neighborhood and the archaeological dig. Figure 2.9 shows the model's temporal output at the city scale: it is the distribution of the number of *historical objects* existing by *temporal entities*. Such a curve reveals the various temporalities through the irregularities: here, the first hop corresponds to the city foundation. The following two hops, even if they may be linked to source bias, correspond to urban restructuration phases.

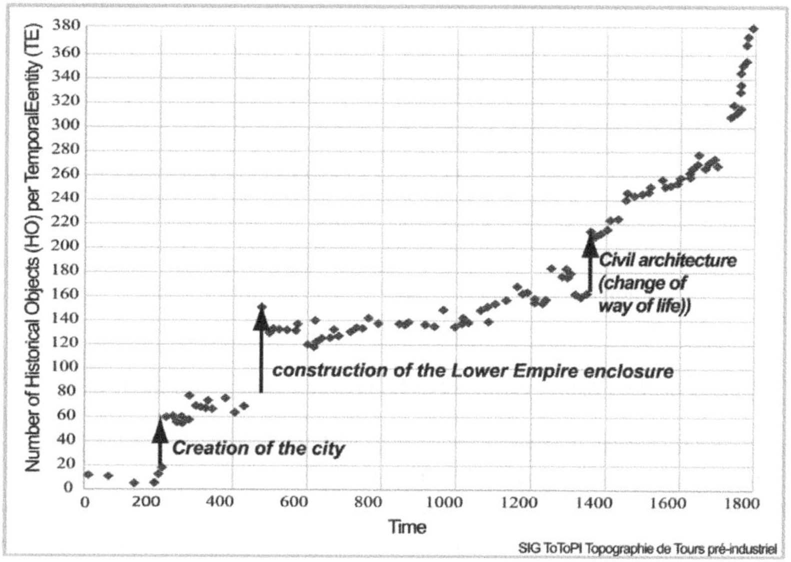

Figure 2.9. *Historical objects' repartition over time (according to [LEF 12])*

Always maintaining the parallel between the spatial dimension and temporal dimension, the authors propose a chronographical representation, through an analogy to a cartographical one. Figure 2.10 is a representation of time and of its characteristics at the scale of the archaeological dig of the amphitheater of the city of Tours. The slicing process corresponds to *temporal entities* that are the periods of stability. These periods are characterized by a certain functional diversity (intensity of gray) and are delimited by borders (vertical bars). The information of the change rate between the two periods can be read on these borders (number of historical objects' disappearances in upper shading and number of historical objects' apparitions in light gray).

It is interesting to see to what extent the representations, which are associated with this example, are far more "time" oriented than "space" oriented: they really serve the temporalities' exploration. This illustrates well the assessment made since

1990s about the way in which conceptual formalism affects operational formalisms: geographers used to question space generate spatio-temporal processes called "space dominant" while historians keep a time-related view and produce spatio-temporal models much more "time dominant" [WAC 99].

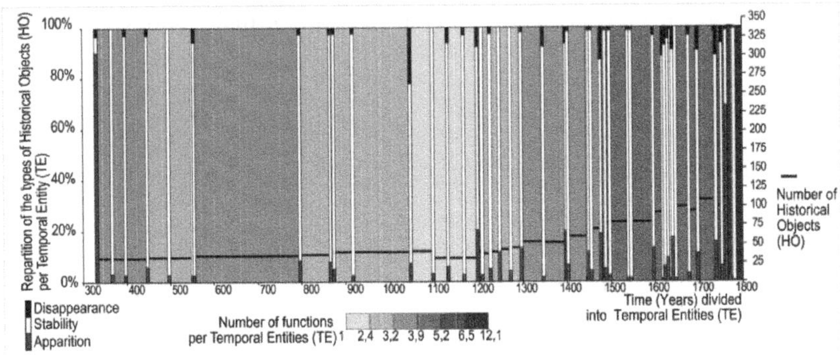

Figure 2.10. *Chronographic (according to [LEF 12])*

2.3.3. *Evaluating the impact of mobile objects on a spatial support*

The methodological challenge of this example is to reveal an approach assessing the possible impacts of mobile objects' behaviors on land use changes. That is Buard's objective [BUA 13]. She develops a methodology to establish relationships between the practices of herbivores herds in terms of consumption of plant resources and of stamping due to their movements, with the transformations of the vegetation cover. This example involves both movement (the movement of herds of herbivores) and places (as spatial regions). The hypothesis to be tested from a thematic point of view is that the vegetation cover is further deteriorating in the places where the herds have been staying, and all the more if these stays have been numerous. From the theoretical point of view, different conceptualizations are possible:

– adopting a continuous field approach and evaluating the properties of each point in space in terms of the animal frequentation intensity and change in the vegetation cover;

– choosing an object approach. The question is then to build up places as spatial portions[13]. On the one hand, there are places of animal frequentation, and on the other hand places that are homogeneous according to the change of their vegetation cover;

13 This approach corresponds to the "categorical coverage" case proposed by Plewe and referred to in Chapter 1, section 1.1.3. A portion of the space is carved out due to its homogeneity relative to several properties of the spatial points.

– developing a crossed approach. This has been the case of Elodie Buard who adopted an object approach for defining the "places of animal frequentation" and a field approach to study the change of the vegetation cover.

We are going to rely on this work to review the four stages around which this chapter is organized.

2.3.3.1. *Step 1: construction of the entities (objects and properties) from the empirical data: identifying the "places of animal frequentation" from GPS readings, and characterizing the change in vegetation cover from satellite images*

At one end of the chain, there are conceptual entities that represent places of places of animal frequentation and the *field* of change of the vegetation cover. At the other end, there is what is observable: – a series of vectors (id, X, Y and t) derived from GPS readings; – satellite images corresponding to two dates t1 and t2. The first step consists of using these observables to build up relevant empirical entities to represent the defined conceptual entities.

The concept of trajectory to build up "places of frequentation" (places of staying and places of transit)

Two complementary inputs can be envisaged to build the necessary entities: – start from the herds of animals and conceptualize them as mobile objects (*endurant entities*) moving in space and staying in different places; – or starting from the movements, considered as trajectories (*perdurant entities*) and focusing on their spatial embedding, on the distances traveled, on the stopping points and on the associated durations. The first point of view corresponds to the phase of collecting information, from GPS collars placed on an animal of each of the *k* followed up herds (during a period of *g* weeks). The second point of view is, however, privileged during the construction of the "places of frequentation". In physics, a trajectory is defined by the "curve described by the center of gravity of a mobile" (Le Robert – French dictionary). Buard [BUA 13] mobilizes the richer meaning developed by Hägerstrand in the framework of time-geography [HAG 70]: the trajectory of a mobile object (in general, an individual) is made up of all his movements and stopping points named "stations" (Figure 2.11). The movements can be used to identify places of transit and the "stations" to identify the places of staying. The series of vectors (id, X, Y and t), collected each hour by GPS (series of events) for each of the observed herds, allows decomposing the trajectories into temporal parts of 1 h and comparing this duration with the distances traveled. This approach allows evaluating the speeds associated with different time intervals, and therefore distinguishing the two types of practices, respectively, corresponding to the travel (greater speed) and sojourn (the "station", corresponding to a very slow speed). The author then identifies the "sojourning places" and "transit places" from the aggregation of information drawn from all the trajectories of different herds (Figure 2.12). The purpose is to search for concentrations of events ("hotspots").

Characterizing change in vegetation cover

The approach here is rather more classic: it is about describing the vegetation cover at two dates t1 and t2 at least, from compatible satellite images, and then to characterize any change at each point. From an operational point of view, the images are processed in the raster mode, and it concerns, therefore, a "pseudo-field" approach (a pixel in the strict sense is a fixed object), the granularity of the observation corresponding to the pixel of the image. The superposition of the information corresponding to the two dates allows identifying the portions of space where the vegetation cover has changed in a similar way, toward an improvement (densification) or a deterioration (weakening) of the vegetation cover.

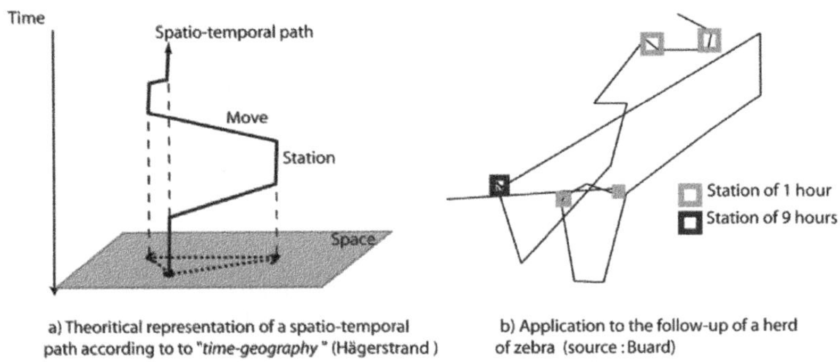

Figure 2.11. *Identifying the movements and "stations" in a trajectory*

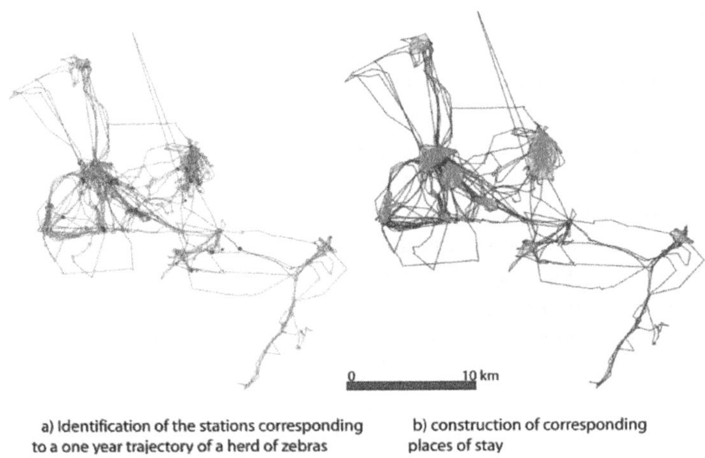

Figure 2.12. *Construction of "places of frequentation" (from [BUA13]). For a color version of the figure, see www.iste.co.uk/mathian/spatiotemporal.zip*

2.3.3.2. *Step 2: to represent and to explore herds' movements and the changes in the vegetation cover*

In this example, the representation of the movement has served as a basis for the construction of the "places of frequentation" as objects. In this sense, the operation is reversed toward to the more common cases where objects are built before considering their trajectories. In other words, focus goes classically from a static point of view to a dynamic point of view. Here in contrast, it is the aggregation of a series of movements that allow building fixed and static objects from the envelopes containing the herd's movements during resting time. Regarding land use, the context is rather classic, the scope of change having been built based on the confrontation of land use at two dates.

Queries can be made from a database (DB), for example: what is the share of pixels recorded as "in deterioration" in spatial portions identified as "staying places" of herds? What is the proportion in the rest of the space? These same questions can be derived from different geographical areas in order to compare them. They can also be derived according to the species of the concerned animal. The same questions may be addressed from a geographical information system (GIS), with a proper visualization.

2.3.3.3. *Step 3: to analyze the relationship between the intensity of animal frequentation and the change in vegetation cover*

The initial question can also be addressed by a statistical approach. The purpose is then a matter of formalizing the information in terms of statistical "individuals" and "variables". In this example, these "individuals" can be either spatial units such as the identified places of frequentation, or pixels. According to the case, different variables can be mobilized, on the one hand, to measure the intensity of the animal frequentation, and on the other hand to evaluate the change in vegetation cover:

– variables characterizing the places of frequentation: the number of sojourns and/or passages of animals herds, the share of different species in this number and the share of space having recorded a degradation of the vegetation cover;

– variables characterizing the pixels: belonging or not to a "place of frequentation", having a degraded vegetation cover or not.

From the methodological point of view, we have to cross the places with regard to the deterioration of their vegetation cover and to the intensity of their frequentation and to look for the possible existence of a systematic relationship between these two phenomena. In the first case, focus is restricted to the places of frequentation, and a correlation analysis will allow measuring the intensity of the relationship between the orders of magnitude characterizing, respectively, animal frequentation and the spatial extent of the degradation of the vegetation cover. In the second case, an assessment can be made to find if there are differences between the pixels frequented by herds and the others in terms of degradation of the vegetation cover.

2.3.3.4. *Step 4: to identify the processes linking animal behaviors and the change in the vegetation cover*

Formulated in this way, it refers to the ecological, ethological and environmentalist domains. A matter that would rather be within the competences of the geographer would be to identify and to model regularities in the relationship between herds' behavior and the nature of the vegetation cover and the topographic context. Such model would, for example, permit to explore the effects of different scenarios on the demography of herds and the parks policy (for example, in terms of the feeding of water points) on the alteration in vegetation cover.

2.3.4. *What factors are behind the emergence and/or the reinforcement of school segregation? Construction of multilevel and multidate information*

By school segregation, we mean the concentration on a certain profile (privileged or on the contrary underprivileged) of pupils in some schools. Works conducted in this area show that two types of factors are at stake to explain the social profile of schools [FRA 08]: on the one hand, it is the reflection of the social profile of the residential area it concerns; on the other hand, it is the consequence of the school choices made by families. Indeed, some families avoid the nearest school, thinking that their child will have better chances of success in a school with better reputation. This reputation can be apprehended by the school's rate of success at an exam or by its more or less privileged social profile. Let us suppose that the objective concerns following three interrelated issues: to better understand how the differentiations of schools' social profiles emerge; to compare the magnitude of these differentiations in a context where there is a "school district mapping"[14] and in a context of free choice of the school for the families; to determine how the schooling contexts that emerge in these different scenarios influence the success of students.

We rely on empirical [FRA 14a, BOU 13] and theoretical works [FRA 14b] conducted at the high school level in Paris and its outskirts.

14 The "school district mapping" is an administrative tool that assigns children to a given school according to their place of residence.

2.3.4.1. *Step 1: construction of the objects and their properties from a multilevel perspective*

The objects to mobilize in order to meet this questioning fall into several levels: the students, the schools and the spatial units (municipalities and school districts mapping). The high schools as well as the spatial units are fixed objects (*endurant* entities), whose attributes and relationships can change over time. A school may appear or disappear, its social composition may evolve relatively to the students who attend it. Regarding students, we follow their characteristics (social, gender and academic success) as well as their mobility from two angles: as a relation between the student's place of residence and his or her place of schooling; as a change of school during the course of the student's studies. Each of these changes at the student level contributes to the change in profile of the school.

2.3.4.2. *Step 2: representing and exploring the pupils' choices of school and the consequences of these choices on the social composition of schools and their evolution*

Figure 2.13 illustrates the spatial and temporal relationships at stake between the student at its place of residence and the school he/she is going to.

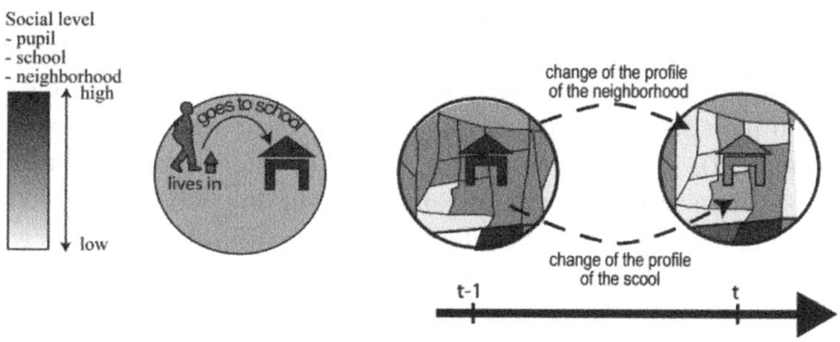

Figure 2.13. *Links between the pupils' movements and higher levels change*

A few examples of questions are given that will guide the exploration of the database and that will involve spatial and temporal dimensions as previously. The special feature of these queries is to link different geographical levels and to mobilize spatial relationships:

– how many students change school during their schooling without changing residence? The students may change school and may change residence. The challenge is to relate the two events;

– what is the evolution of the number of pupils attending a school outside of the district they belong to? Students move, the school district mapping changes, but this question is specifically about the evolution of the relationship between the pupils and frequented school. The result depends on the relative localization of each pupil and her/his school;

– how many schools have an underprivileged social profile? Such a query consists of selecting schools according to a summary of the students who frequent it. For instance, selection can be done using the % of students in the "underprivileged" social category in comparison with the same percentage at the level of the region;

– what are the schools whose share of underprivileged pupils has increased during the period? This query requires an analysis of the evolution of the targeted indicator's trajectory.

Queries can multiply. It can be seen that the complexity increases quickly when the queries question evolutions that mobilize composition or spatial relationships, which are themselves evolving.

2.3.4.3. *Step 3: to analyze the relationships between school success the students' and the schools' characteristics*

Simple methods of statistical bivariate analyzes (correlation, variance analysis, chi2 testing and following the qualitative or quantitative nature of the variables at stake) allow examining the relationships between differentials in school success and social inequalities. At the student level, the point is to test whether there is a significant relationship between their social category and success (measured, for example, from their success or not at the brevet (French high school exam), or from the grade obtained). At the school level, the main interest lies in the relationship between the success rate and the social profile of the school (derived from the proportion of pupils whose parents belong to certain well-off social categories or on the contrary underprivileged, or whether from the first component of a principal component analysis on the % associated with the different social categories).

The next step consists of developing multivariate models (multiple regression or logit models) that help to explain the "success" in function of several explanatory variables. At the student level, a logit model allows, for example, showing that the social category of the student and the social profile of the school she/he attends are similarly discriminating in the public sector, whereas the social profile of the school is more discriminant in the private sector [FRA 14a]. At the school level, a multiple regression model allows showing that *everything being equal* with regard to the social profile, the interschool differentials in terms of success at the brevet (national exam), are related to the importance of the migratory turbulence of the school (i.e. the importance of incoming and outgoing flows of pupils during high school years). The analysis of the residual map of such a model allows highlighting the local specificities, with varyingly important success rates than what was expected from the model. Figure 2.14 shows the results for public sector schools as an example. It allows assessing the spatial variability of results at the regional level and identifying the concentrations of schools with similar values. That of the North-West of the region is rather significant, underlining an area where almost all schools record lower grades to what would be expected from the model. The picture that can be seen is one of an area with strong social discontinuity within the region, whose effects are underestimated by the model. Undoubtedly, other effects which are not easily measurable can be added to this such as the leakage of good students at the school level, the inequalities of school funding and possible inequalities in grading (diploma grading maintained at an infraregional scale).

2.3.4.4. *Step 4: to identify the processes leading to a varyingly significant school segregation*

The development of a simulation model is useful in order to identify the processes in play in school segregation. It will permit us to differentiate between the respective impact of the unequal distribution of the population according to its social profile at the place of residence, of the practices of schools' heads with regard to the selection of pupils; and finally, that of the practices of students' families in terms of school choice. The multiagent system formalism allows, therefore, creating an artificial world populated with students and schools led by school heads. The objective is then to determine, in order to test them, the choice rules of *student-agents* and the selection policies of *school heads-agents* [FRA 14b]. From this model, different scenarios can be implemented (for example, more or less strict school district mapping policy), the benefit being to be able to evaluate the final configurations by comparison.

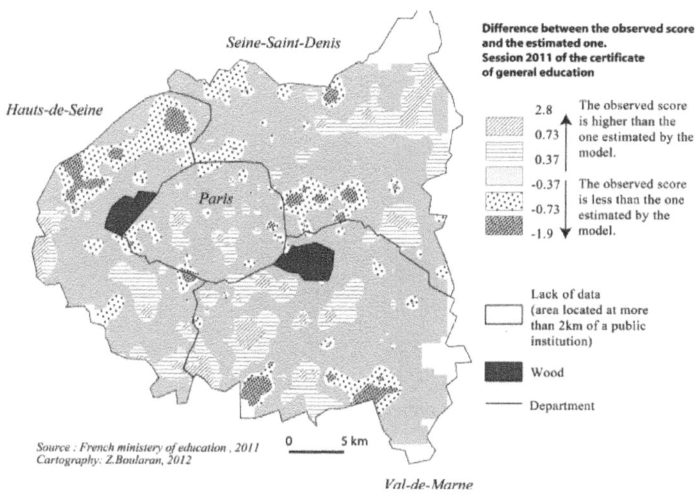

Figure 2.14. *Spatial organization of residuals between the observed and modeled grade in Paris and close outskirts*

2.4. Conclusion

In these examples, we have resolutely assumed a "semantic" positioning rather than a "computer science" oriented one. We have tried to enlighten the developments presented with respect to different categories introduced in the Chapter 1. They are not always explicitly expressed, but clues are scattered all along the text. Priority was given to addressing in a comprehensive and detailed manner the different stages of the processing chain, and the choices which are associated.

Indeed, the objective of this chapter was to enlighten and illustrate the stages of an analysis process in an interdisciplinary context, where a number of points of view coexist (geomatician, computer scientist, geographer, statistician, modeler, etc.). Different conceptual frame, methods and tools are chosen according to disciplinary practices and cultures. This does not exclude the fact that there is a commitment to respond to the same question. It is the interpretation of the question, through the specific methodological competences, that makes a difference: the disciplines are rare where only the formal aspects filter the question and decompose it in a single frame of reference, without interference from technical systems.

Chapters 3 and 4 deepen two of the four challenges presented in this chapter: that of the analysis of the evolutions of statistical and spatial relationships and that of the exploration of processes underlying change. The emphasis will be placed on how time and space are taken into consideration to account for social and environmental systems' dynamics. Chapter 3 concerns changes in spatial configurations and evolution of the spatial entities making up these systems. In chapter 4, the focus will be on the spatial interactions and their role in driving these systems' dynamics.

3

Analyzing Spatio-temporal Data: Empirical and Statistical Approaches

This chapter is dedicated to the presentation of spatio-temporal approaches that rely on statistical methods (descriptive or explanatory) to analyze changes that occur over the geographical space and in spatial systems. It is not a manual of spatio-temporal statistics. The objective is instead to present an approach that relies on methods commonly used in the social sciences, and just like in the previous chapters, to discuss the choices made in function of the manner in which space and time are mobilized. We will therefore assume that the basic statistical methods are known. We will focus on the approaches that allow, through different types of combinations and coupling of methods, the highlighting of the main trends of change, both at the level of the entities involved and that of the relationships that they maintain as well as the configurations that they draw.

Section 3.1 presents the disciplinary landscape in which the field of spatial analysis has evolved; the evolution being marked by the statistical methods and the progress of computer science. This landscape and the interactions between its various components, including statistics and computer science, have left their mark on the practices in terms of data processing. These practices are based on a certain "hybridization" of methods and on the central place accorded to the data as well as the meaning that is given to them. We then introduce a duality of objectives for the spatial analysis of spatio-temporal data. This approach consists of following over time spatial patterns on the one hand and following localized entities in order to analyze the spatial organization of their evolutions on the other hand. This duality will serve as a main thread to the developments presented in sections 3.2 and 3.3.

3.1. Statistical data and spatio-temporal analysis

The practices in spatial analysis have been strongly influenced by the different methodological developments in statistics and in computer science. On the basis of these disciplinary fields, geographers have devoted their time to developing formalisms adapted to their questions, with a particular attention to the taking into account of geographical space. We sketch out here a few of those stages and then go back to the layout of the data.

3.1.1. *Spatial analysis evolution in a context influenced by statistics and computer science*

As early as 1960, in the context of "the quantitative revolution", Brian Berry [BER 64] proposed in a seminal article a system of generic representation that allows the integrating of the variety of empirical approaches of regional science. Starting from the "geographical fact" that he describes as being either the observation of the same phenomenon in several places (topical approach), or several observations in the same place (regional approach), he proposes the concept of a "geographical matrix" (which is also known as a "spatial information matrix"). This concept allows these two points of view to be integrated. The rows of this matrix designate different measurable characteristics and the columns designate the places. At the intersection of a row and a column, the observed value is found for a certain characteristic in a given place. The geographical matrix is a representation system for the analysis of spatial variation (variability over a row) and that of spatial associations (spatial variability over a column). For Berry, this representation is compatible with and supports a systemic point of view: all the described locations are interact and make up a spatial system. The geographical matrix allows the evolution of such a system to be conceptualizes by taking a third dimension into account, the temporal dimension. A move is then made from variations and spatial associations to variations and temporal associations (Figure 3.1). The same components are found in this formalism as in the one developed by Peuquet [PEU 84] presented in the previous chapters.

Berry's proposals may seem very "modern" nowadays but they were inserted into a context where automatic calculation was only just beginning to emerge. Quantitative geography (and spatial analysis) has benefited from the strong progress that the two disciplinary fields on which it relies have experienced:

– the statistics for all the methodological aspects and data processing;

– computer science for all aspects related to geographic information and data modeling.

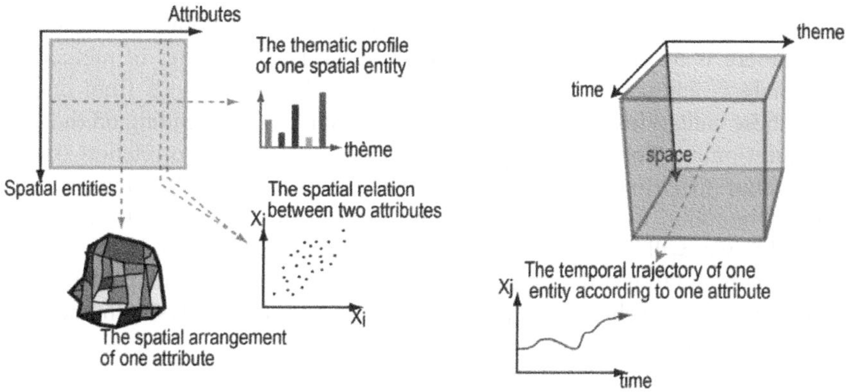

Figure 3.1. *From Berry's spatial information matrix to the exploration of spatio-temporal data [BER 64]*

The progress made in these two areas has been transferred to geography at different times, with varied issues (1960 for the development of quantitative methods and 1980 for that of geographical information systems). If spatial analysis has been particularly influenced by the developments of these two areas, this is mainly because they impact simultaneously on the methods that it uses and on the information which it works upon. This impact is all the more important given the increase in the availability of information, its diversity and the richness of its granularity. A detour describing this landscape is useful to place our positioning.

With regards to the statistical methods, which are concerned here, there have been numerous discussions between statisticians and geographers, highlighting differences in points of view. These interactions, which are extremely rich ([CLI 73, ANS 95, GRI 10]), have resulted today in a set of methods upon which spatial analysis is based. According to O'Sullivan and Unwin [OSU 10], three types of methods can be distinguished:

– *analysis of geographical data*, which is based on descriptive and exploratory statistical methods to identify statistical relationships (in the sense of Berry's "associations") between various spatial phenomena [CHA 87, LEB 06, DUM 11];

– *spatial statistics* which have been developed to test the significance of the random or organized characteristic of a spatial configuration [CRE 93, ANS 95, GRI 91, FOT 00, RIP 05, DUB 14);

– *spatial modeling* which relies on models taking into account a spatial dimension, such as Reilly's model to define catchment areas, or the gravity model to predict flows between places [HAG 65, TAY 77, PUM 97].

This categorization reflects historical and disciplinary points of view, but in practice, there is no compartmentalization between these families of methods. The spatial analysis approaches combine, as a matter of fact, methods from one or an other of these categories, and can give them different roles depending on the context of the questioning. For example, a coefficient of spatial autocorrelation (indicator falling within spatial statistics) can be utilized as a variable describing the intra-urban configuration of a set of cities in a process of geographical data analysis. Conversely, we will be able to mobilize the nearest neighbor method (spatial statistics methods) to test the spatial dependency of a synthetic index characterizing the social profile of schools computed from a multivariate analysis (method within the scope of factorial analysis).

The analysis of geographical data has benefited in particular from the contributions of descriptive analysis methods such as multivariate data analysis[1] [BEN 80, LEB 06] and the exploratory analysis emanating from the current of "exploratory data analysis" devised by Tukey [TUK 77]. Unlike the classical confirmatory approach in statistics, which consists of testing a hypothesis that is made explicit *a priori*, the exploratory approach proposes to use a set of methods and representations deriving from descriptive statistics. This is in order to identify relationships and tendencies that will allow us in a second step to lead to hypotheses about the state and the functioning of the system that is of interest. These hypotheses will in turn lead to new and more targeted explorations and, step by step, will allow new knowledge to be highlighted about the phenomenon being studied. This approach supports the abductive approach referred to in Chapter 2, which involves identifying *a priori* any surprising fact in order to follow it up and explore it from different angles. In the field of geographical data, this type of approach has been developed by synchronizing maps and graphs in interactive environments [UNW 90, UNW 94, BAN 01]: the explorations are conducted in a very visual manner from a selection-query through maps or graphs, applying a specific analysis on the selection.

Today, with the increasing volumes of available data, these approaches are integrated in the emerging field of "data mining and knowledge discovery" that finds a geographical specification [WAC 02, MIL 09]. The new masses of data are most often not derived from explicit collection protocols and are therefore imperfect and not fit to be used in the often very restrictive conditions that the conventional statistical models presuppose. Rather than testing *a priori* hypotheses, these approaches propose sequences mobilizing varied analysis methods in order to reduce and/or enhance the information and detect patterns of interest: automatically

1 Particularly in France where Benzecri's school has developed an original method, this method was specifically adapted to qualitative data (correspondence analysis or factorial analysis of correspondence), which has allowed very strong diffusion in the areas of the social sciences (sociology, geography, psychology, etc.), facilitated by the development of free software and training.

selection and layout of the data (methods that consist essentially of cleaning and estimating the missing data); enhancement (by external data integration); reduction (of dimensions analyzed by factorial analysis for example, or by a value estimated by a statistical model). Furthermore, whereas the dependency between neighboring observations in space and the heterogeneity of spatial processes involved made the geographical data unfit for classical statistics methods, a number of specific methods have been developed that integrate these characteristics and use them to give a different insight to associations and spatial distributions [FOR 02, GUO 09, MAT 06]. All these developments have led to making the specificity of geographical data visible and to giving its own status to the spatial dimension taken into account in analyses.

Nowadays, more than parallel approaches, the simple explorations, as well as the explorations that we will qualify as abductive, to emphasize the going back and forth approach mentioned above, and the confirmatory analyses, make up a set of complementary approaches with sometimes very thin borders. Our position is at the interface of these different approaches, combining different types of statistical analyses to identify and describe the relationships embedded in the geographical space and evolving over time. This could be the evolution of the correlation between the density and the distance to the center within a metropolis, or the evolution of the differences in the school performances of students according to gender, social category and localization of attended school. It could also be the evolution of flows of commuters between the municipalities of an urban area according to the evolution of employment and the transport network. In all three cases, the purpose is to understand a change in a relationship. The first case concerns the relationship between variables characterizing spatial units, the second case concerns the relationship between variables measured at the level of pupils and the third case concerns the direct relationships between spatial units.

These approaches are conducted putting the reflection on the to-and-fro between data types and types of analyses. This to-and-fro is facilitated on the technical level by the development of new environments, but in the current context of multiplication of information sources, there is a risk of shifting from "the observation of the phenomenon to the observation of the data"[2]. Our choice is, on the contrary, to favor an approach that gives a central place to the studied phenomenon and to the entities associated with it. The question of the spatial and temporal resolution of the observations that are linked to it benefit from being processed at the conception stage of the objects of interest, as has already been mentioned in Chapter 2; even if it is often raised at the time of the implementation of analyses.

2 R. Laurini, EVS seminar February 17, 2014 Lyon: "*Are there fundamental principles in Geographic Information Science?*" with N. Chrisman.

3.1.2. *Different ways to take time and space into account to analyze spatio-temporal processes*

Statistical analyses relate to a data table, describing a statistical population (rows) according to a certain number of characteristics (columns). The population is composed of homogeneous statistical individuals from a semantic point of view. The notion of a statistical individual corresponds to the concept of an object developed in Chapter 1. It is the basic constituent of the studied set and it is in this sense regarded as indivisible: the students of an academy, the shops of a city, the plots of a forest but also the cities of more than 10,000 inhabitants of a country, etc. In its most generic form, a statistical table describes a set of individuals by a certain number of variables (also called characteristics or attributes). When it comes to describeing a spatial phenomenon in time, it is necessary to identify these two dimensions, the spatial and the temporal ones, in the data. Further in the text, we give detail about the different manners to formalize time and space in this type of table.

3.1.2.1. *From the statistical table to the geographical information matrix: "support" space and "active" space*

With the explosion of observation means and the easy access to individual data, the questions relating to spatial configurations and processes linked to human activities rely on data constructed from observations of increasingly finer spatial resolution, going as far as to the located person. Further in the chapter, the presented methods are more specifically applied to tables of geographical information.

Berry's matrix of geographical information is a particular case of a table of data where the statistical individuals are places (or geolocalized object). It is initially by this specific status of individuals that space intervenes in the analysis. The spatial dimension is, however, not explicit: everything happens as if the places were a statistical population without structure while what is happening in a given place is precisely not independent of what is happening in neighboring places[3] [TOB 70]. A simple regression between the population and the average income of the municipalities of a region, for example, does not give any *a priori* role to space. The spatial structure may be identified *a posteriori*, by mapping for example the residuals of such a regression. In this case, we will qualify the space as "support" . Only the issues related to the spatial variability (heterogeneity of a row) of the phenomena being studied can be processed and not those related to the spatial ordering of the spatial units involved. Figure 3.2 illustrates the difference between statistical heterogeneity and spatial heterogeneity.

3 According to Tobler, "Everything is related to everything else, but near things are more related than distant things".

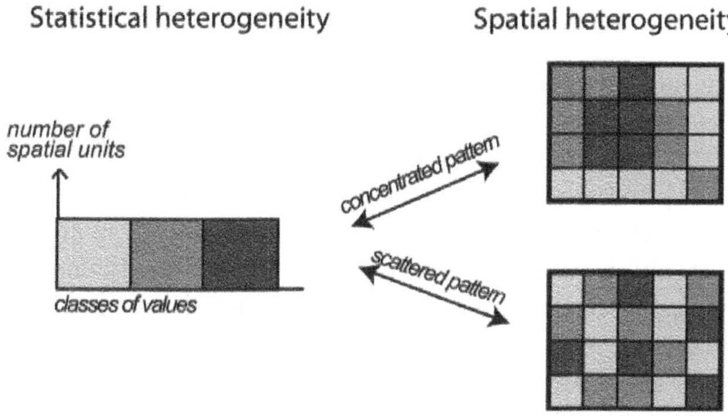

Figure 3.2. *Statistical heterogeneity and spatial heterogeneity*

Spatial analysis gives, by definition[4], a central place to space in the analysis of the phenomena of interest. The simplest method consists of building spatial descriptors and formalizing that way the spatial structure of the phenomenon being studied: the most classic example, which is at the basis of numerous spatial models relies on the hypothesis that distance is a structuring element (distance to the center, distance to the facilities, distance to the network, etc.). However, we can also imagine describing a place by a characteristic *and* this same characteristic measured in its neighborhood: if the example of the average income of the inhabitants of a municipality is recalled, tests can be carried out in order to measure the effect on a municipality of average wealth to have wealthy or poor municipalities surrounding it. Another way of integrating the role of space is to situate the analysis subject in the meshes of the zonings in which it is located: the academy for a high school, the region for a city, the climate zone for an agricultural plot. These higher level spatial entities can be mobilized as a relationship of belonging or through descriptors characterizing the higher level spatial unit (a rate of school success for the academy, the GDP for the region, the duration of the episode of drought for the climate zone). These relationships between zonings are also a form of expression of the effects of the neighborhood [MAT 07]. In these different cases, space is said to be "active" because its structure (distance, neighborhood, mesh hierarchy) plays a role in the analysis and appears to make explicit the differentiations existing in space for the phenomenon of interest. Therefore, the notion of "situation" is integrated in the objects' attributes, which, as other attributes, can evolve over time. This type of statistical table is undoubtedly the one which has been the most studied in the field of geographical data analysis.

4 In this context, Charre's definition of spatial analysis is particularly eloquent: spatial analysis consists of "searching for, in the characters of spatial units, what refers to their geographical position, in particular relative, which makes it necessary to model the spatial structure" [CHA 95].

3.1.2.2. *"Unfolded" time and "folded" time*

When Berry adds time to its "spatial information matrix", this involves three dimensions. It is then interesting to recall the analogy between spatial dimension and temporal dimension discussed at length in Chapter 1. According to the terminology used by [GAL 04], the temporal dimension can be regarded as a field where each point can be informed by the state of the studied system: these are for instance the dates of the population census to describe the population dynamics of cities; the dates of school start to describe the social composition of high schools; the dates of the satellite images to describe land use in a region. Galton qualifies this time conceptualization as "history", as opposed to "chronicles" where the temporal dimension is made explicit in terms of events. Two cases can thus be differentiated:

– "histories": the X variable being studied is measurable *for all statistical individuals*, on each of the dates. The variable X is thus stamped (dated) and replicated in time, $(X_t \ldots X_{tk})$. It should be noted that in a table hence formed (just as for space) order exists only through the indices. Some specialized statistical analyses will take this order into account (methods of time series analysis) but in most applications, the user will interpret this order *a posteriori*. Berry's table corresponds to this situation in the case of a description of the places by several variables. $(X_{1,t1} \ldots X_{1,tk}, X_{2,t1} \ldots X_{2,tk, \ldots})$. Each temporal plan therefore describes a "history" that characterizes the state of the system at this date;

– "chronicles": the sequence of events *differs from one statistical individual to the other* (Ms. H gets married, divorces, remarries and then becomes widowed, while Ms. S remains single; the municipality X merges with another, joins an inter-municipal union, and then is integrated into an urban unit). Some specialized methods allow such trajectories (*event-history analysis*) to be analyzed directly. In the other cases, in order to integrate these different temporalities in a classic statistical table, it will be necessary to build indicators summarizing the chronicle. For example, it will be possible to calculate the number of events in the period, the date or the duration of an event or the average duration between two events. In this case, the indicator is not stamped but it "contains" time.

It is possible to go from a "history" formalization to a "chronicle" formalization by defining events (for example, the date at which the population exceeds a certain threshold), and vice versa for a "chronicle" formalization to a "history" formalization. This can be achieved by associating, for example, a state with each event: the state of "existence" is, for example, related to the "emergence" event (of a shop for example) and the state of "absence" with the "disappearance" event. These states may then be projected over the time line over the considered period.

These different time formalizations suggest an analogy with the differentiation that we have introduced between a "support" space and an "active" space. Here we propose the term “unfolded time” when the time is made explicit by dates (analogy with the “support” space) and that of “folded time” (analogy with the “active” space) when time is taken into account via indicators summarizing the duration and the events. The first case refers to Galton’s “histories”, where time “carries” the information about the system’s status, the second refers to the temporal summaries in a “chronicle” perspective. According to the question, one or the other approach will be adopted in the applications. If recalling the example of the municipalities’ populations, it is therefore possible to work directly on the table of populations at each date and to build demographic trajectories or to build indicators summarizing the characteristics of change. This may be the variation in population over the period. In this case, a differential of quantity between two dates is considered. This can also be about a date or a period corresponding to an event which has some meaning for the question: the date of the maximum population, the period during which the average annual rate of growth has exceeded 1% per year, for example. The summary indicator is a measure of time and can be processed as a quantitative or ordinal variable. This second type of formalism is very often used because it allows formalizing temporalities *a priori*.

Recalling the example of the geographical information matrix, Berry proposes a table of data where time is “unfolded” and the space is of the “support” kind. The formalization of an “active space” and a “folded time” complements this approach.

3.1.3. *From entities’ dynamics to that of organizations: what is it that changes and relative to what frame of reference?*

Data formalization depends of course on the question that is raised, and on the hypotheses available. Being interested in spatial dynamics, it is important to identify the *level* to which the question precisely refers. Two families of questions can be distinguished:

– The questions about *change in spatial organization*: the observed spatial entities then correspond to a means of observation and are simply representative of the observation granularity that is adopted, either this entity has a meaning for the study (municipality, archaeological settlement, shop, etc.) or not (pixel, measurement points). The interest is focused on the description and identification of changes at the level of the whole set of entities considered. It may be the evolution of the spatial organization which is targeted: what is for example the evolution of the poverty configuration in this region? What change in land use has been caused by the demographical dynamics of this city? It could also be the hierarchical organization that is in focus, since it is often associated with these spatial recompositions: what hierarchization recorded the settlement system between the 1st and the 2nd Century for example?

– The questions about *the dynamics of spatial entities* themselves: these entities then have a "meaning", and a rather "object" oriented approach is then involved. The analysis will focus on the evolution of these objects: the point will be to look for similarities between these evolutions, as well as identifying types, and assessing potential gaps between the changes of the various objects. It will be possible to focus *a posteriori* on the spatial configuration of these evolutions: How are the demographic trajectories of cities spatialized? Is there a regionalization of the municipalities' poverty evolution? Where are the parcels that have resisted the urbanization front?

Here again, it is interesting to draw a parallel with the points of view discussed in Chapter 1. Galton [GAL 04] for example introduced two possible decompositions for the spatio-temporal "field". It can be seen as a sequence of temporal crosscuts (*snapshots*), or as a juxtaposition of temporal trajectories of spatial entities (Figure 3.3). Parallel to this distinction of an epistemological nature, the statistical data will be laid out differently and/or different analyses will be mobilized depending on the objective pursued.

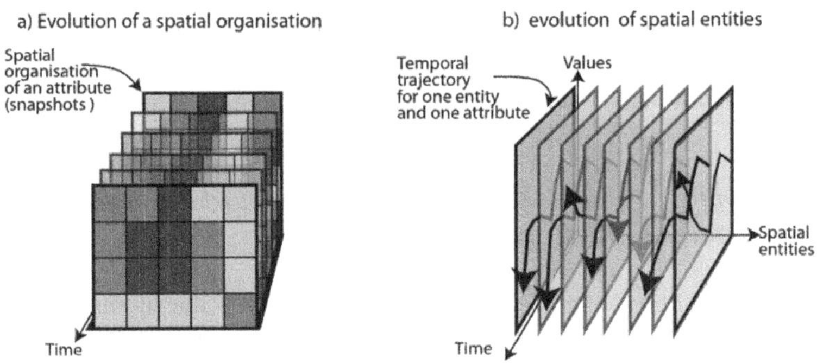

Figure 3.3. *Spatial evolution: two points of view*

Moreover, in an evolution analysis, whether it is about a spatial organization or spatial entities, the issue of the reference arises: in relation to what is change being observed? Any diachronic study is faced with this question, whose answer is often not unique and depends on the point of view adopted to define the referential frame. Let us take the example of a system whose elements are described by a single variable, such as the cities of a region described by their resident populations at the different censuses since 1800. The description of their absolute evolution, which is through the number of inhabitants, or relative, in relation to the urban system that they are part of, gives two different images of the urban growth (Figure 3.4). Let us take the example of the Rhone valley urban system (France):

– in absolute terms, the evolution of the urban system is characterized by an overall growth, all the cities registering a growth of their population;

– in relative terms, i.e. by relating, for example, the population of each city to the population of the largest city, the representation underlines a growth concentration process at the top of the urban hierarchy (Figure 3.4(a)).

The evolution of the populations of Nimes and Montpellier offers another example (Figure 3.4(b)). The representation of the populations in relative values (compared to the total of the urban population of the region) allows the discontinuity occurring since the 1960s between the population trajectories of the two cities to be highlighted.

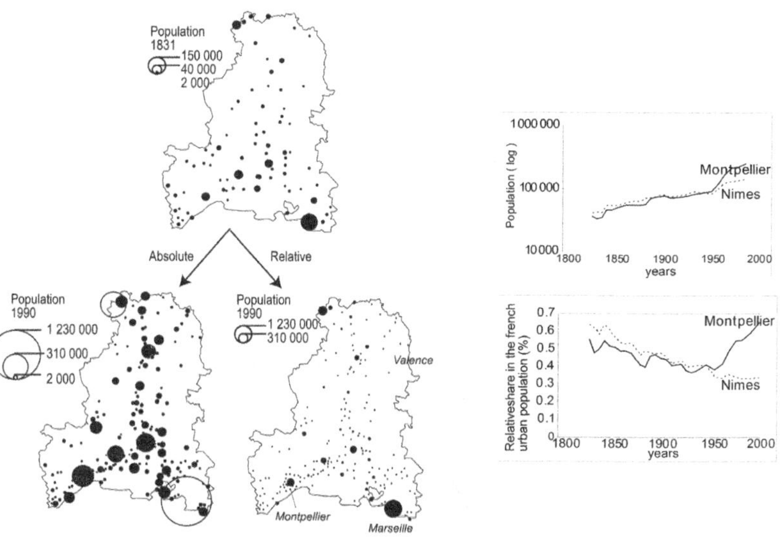

a) Evolution of cities'population in the low Rhone valley (Southern France).

b) Comparative evolution of the population of Nîmes and Montpellier

Figure 3.4. *The issue of the referential in the change analysis (source [ARC 98] (revised L. Lizzi))*

The question of the referential may also arise at the level of the involved methods. Let us suppose that the interest is about the follow-up over time of a spatial organization. In a "snapshot"-based approach, the same static model is used at each date as a grid of interpretation of the spatial configuration (calculation of a spatial autocorrelation coefficient at each date, for example). Each result will be an element of the trajectory describing the evolution of the structure (here a spatial autocorrelation coefficient trajectory). If the method depends on parameters (in the case of spatial autocorrelation, this would be the adjacency graph for example) the question may then arise about the evolution of these parameters over time, and thus about the reference frame associated with the method upon which the interpretation will depend. It is also typically the case of multivariate factorial analyses made at several dates: while the shares of the variance explained by the first factor at each date can, for example, be compared, the positions of an entity on the first factor computed for different dates cannot be directly compared. Indeed, these analyses are describing differentiations according to a specific reference frame at each date.

The rest of this chapter will be dedicated to applications of spatio-temporal analyses. The distinction between an input directly centered upon the evolution of the spatial organization *versus* an input centered on the evolution of entities, is used as a support for the adopted plan. Section 3.2 is dedicated to the first of these inputs and to the steps to understand the evolution of a spatial system's structure. We will distinguish here between two types of approach:

– those consisting of calculating a time series of indicators summarizing the spatial organization (for example a center-periphery gradient or the spatial autocorrelation of a phenomenon) (section 3.2.1);

– those based on the follow-up in time of spatial associations (correlation between variables describing spatial entities, for example, or factorial analyzes in the multidimensional case) (section 3.2.2).

Section 3.3 presents the approaches centered on the evolutions of the spatial entities themselves. The point is to describe and explain (in the statistical sense) change at the level of these entities (by integrating a number of explanatory factors and various temporalities) and to represent the trajectories corresponding to their evolution (section 3.3).

3.2. Following the evolution of the structure of spatial systems

The examples presented in this section give an account of the evolution of the structure of a spatial system, whether concerning its spatial configuration, another kind of structure, hierarchical, for example, or even the relations' structure between the variables characterizing the entities of the system considered.

3.2.1. *Describing the evolution of space and hierarchical organizations*

3.2.1.1. *The indicators measuring spatial organization*

The objective is to implement methods to give account of the different "states" of the spatial organization of the system studied in order to identify and measure the different aspects of change. In some cases, very simple indicators allow us to show spatial change, in others, it will be necessary to mobilize the indicators built from spatial statistics. For example, "the distance to the nearest neighbor or to the *k*th nearest neighbor", typically used in probabilistic approaches of spatial configurations [RIP 05], may be used to describe the degree of concentration in a set of points at multiple dates. All these measures thus constitute a summary of the "history" of the whole spatial organization. Let us take the example of a set of archaeological settlements stamped following a time granularity at the century level in a region in the South of France (the Lunellois and the Vaunage [FAV 98]). These settlements are described by a certain number of indicators characterizing the level and the activity of the settlement as well as the function of the settlement. The analysis of the coevolution of the number of occupied sites and of their areas provides the first image of the evolution of the settlements' structuring and allows the different occupation phases of the territory to be identified. A deployment of the settlement system can be observed at the beginning of the period with a very strong increase in the number of settlements with small areas. This was followed by a retraction, the settlements being concentrated on a smaller number of larger settlements with a more significant area [NUN 06] (Figure 3.5(a)). Local indicators allow a full account to be given in parallel to these evolutions from a spatial point of view. Figure 3.5(b) shows the evolutions of the averages of the "distance to the nearest settlement" and of the "number of neighboring settlements within less than 3 km". These indicators, measured for each settlement occupied during the period being considered, are two complementary ways to give insights about the local situations. They allow the spatial inscription of the phenomenon to be highlighted. The change of the average distance between settlements from 500 m to more than 1 km between the 1st and 5th Century illustrates the emergence of a new spatial organization of the settlement system, with a more regular coverage of all the territory.

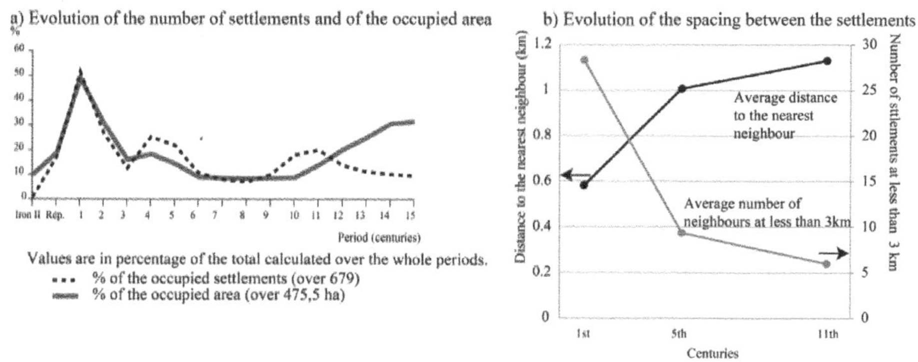

Figure 3.5. *Statistical description of the evolution of a point pattern (source [FAV 98])*

The follow-up of position and dispersion indicators over time are also simple ways to identify changes in the organization of a spatial system. Two examples are given here:

– The gravity center of a set of points allows following the tendency of its spatial distribution. Thus, in the framework of the analysis of the evolution of the postal roads in France between 1534 and 1833, the spatial "trajectory" of the gravity center of the posthouses allows illustrating the manner in which this network of postal roads has evolved (Figure 3.6(a)). The move of the gravity center toward the West and then to the North illustrates the fact that the network has heavily become denser, first around Lyon, then in the North of Paris and this happened very rapidly between 1530 and 1650. This trend stopped around 1750, and the movement of the gravity center toward the south illustrates a catching up phase with regards to the South that continues until 1810. Between the last two dates the center of gravity remains fixed, illustrating that from this point the progression of the network is done in a balanced manner over the whole of the territory [BRE 05].

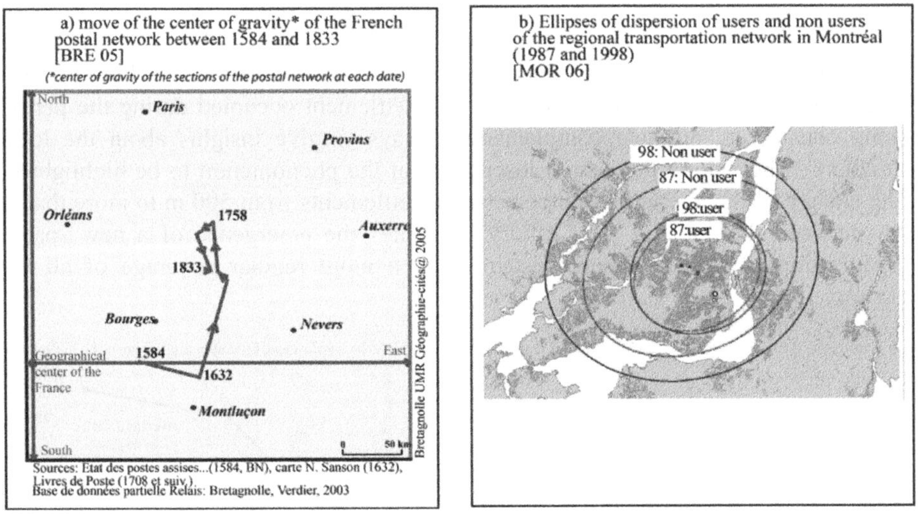

Figure 3.6. *Changes at the spatial organization level*

– The "dispersion ellipse", which is an indicator of dispersion around the gravity center, can also be used to describe the evolution of a punctual phenomenon over time. It concerns a geometrical shape, from which certain characteristics can be reused (for example the surface, the maximum radius, or the ratio between the two radii). Figure 3.6(b) is a good example. The ellipses correspond to the residence envelope of two categories of persons, according to whether or not they use the regional network of public transport in Montreal at two dates [MOR 06]. The set is composed of the persons' residences. By comparison, the dispersion of residents taking public transportation does not evolve between 1987 and 1998, while those of individuals not taking any transport evolved considerably. This type of insight, essentially spatial, allows us to question the adequacy of the offer of transportation in a context of deployment of peri-urban places of residence for persons working in Montreal.

The indicators used to describe the transformation of a spatial organization can also be the result of statistical modeling. If we consider, for example, the gravity model[5] (presented in Chapter 2) that consists of modeling the intensity of exchanges between places in proportion to the masses of places and in inverse proportion to the distance between them [HAG 65, TAY 77]. It can be used diachronically to show, for example, the evolution of the parameter associated with the distance, called the distance friction, at several dates. Thus, if it is used to model the commuting characterizing an urban area, the model will be expressed as:

$$F_{i,i,t}=k(t)\ M_{it}.M_{i't}/D^{b(t)}_{ii't}$$

where $F_{i,i't}$ is the value of the commuting flow between the municipalities i and i' at the date t, M_{it} and $M_{i't}$ are, respectively, the active population residing in the municipality i and the jobs of the municipality i' at the date t, and $D_{ii'}$ is the distance separating the two municipalities. All the quantities are likely to evolve in time. The two parameters k(t) and b(t) represent the mobility rate (the share of effective exchanges among the exchanges theoretically possible) and the distance friction (the dissuasive effect of the distance). The variation of the distance friction during time b(t1), b(t2), b(t3), etc. gives a new image of the exchanges' follow-up over time: it allows us to assess the role of the distance in the change and approach that of the range of a labor center's attraction area. A similar approach can be developed from Clark's model (also presented in Chapter 2), which allows the decrease of cities' density to be modeled as a function of the distance to the center [MAT 00]. In this case, it is the population that is modeled for a set of municipalities in the region of the Rhone valley in France (around Valence):

$$P(i,t) = a_t \exp(-b_t x_i)$$

5 In its most general form, the gravity model is written: $Fi,i'=k\ Mi.Mi'/D^{b}_{ii'}$ where Fi,i' is the value of the exchange between the places i and i', Mi (respectively, Mi') are the weights associated with each of the places assessing the emitting capacity of i (respectively, receiving i') and Dii' is the distance between the two places.

where P(i, t) is the residential population in the municipality i, x_i is the distance between the municipality i and the center of the Valence and t is the observation date (the distance is measured by the Euclidean distance and is therefore supposed constant in time). The two parameters "a" and "b" represent, respectively, the density in the center and the dissuasive effect of the distance. They evolve over time, and the b(t) parameter of the model can be used as an indicator of the shape of the densification process around the center whereas the sequence of the values of a(t) over the period is associated to the growth of population in the center. The combination of the two illustrates the different components and timescale of the "spreading" process associated: first it is possible to observe a strong growth of the urban population at the center of the region and in its surroundings. Between 1850 and 1950 the growth is very low, and after 1950 growth it picks up due to rural exodus: the center becomes denser whereas surrounding places emptied (Figure 3.7)

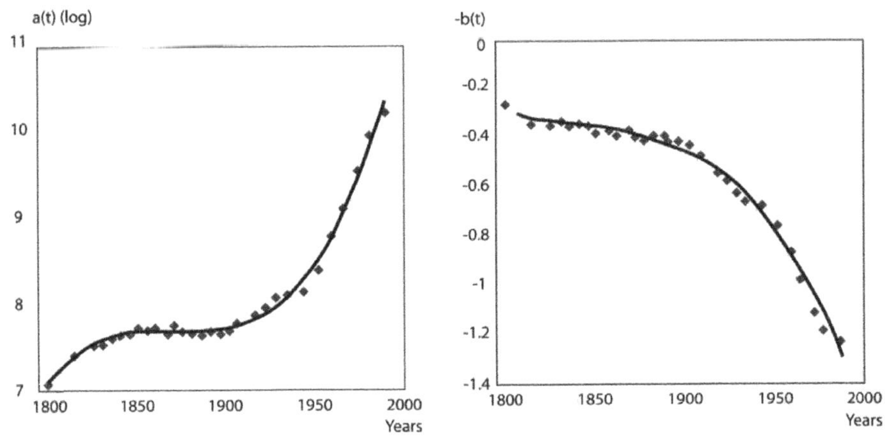

Figure 3.7. *Spatial model and follow-up over time of a structure (according to [MAT 00])*

3.2.1.2. *The rank-size relationship as an indicator of the hierarchical organization of a system of cities*

The rank-size[6] distribution reflects regularity in the relationship between the rank of a city and the number of inhabitants who live there. The slope of the line

6 The rank-size distribution expresses the regularity of the relation between the cities' sizes and their rank in the urban hierarchy (one refers to Zipf's law (1949) [ZIP 49] when the slope of the line is equal to 1, and to rank-size rule otherwise). The empirical applications concern both contemporary and historical periods and more distant periods. In archaeology, this framework is often used to characterize settlement systems whose settlements are not cities, but much smaller agglomerations.

associated with this relationship is an indicator of the degree of hierarchization of the settlement system. Many researchers have focused on the universal character of this relationship and there are numerous empirical works showing its adequacy [BAT 01, MOR 93, PUM 06, GUE 95, PEA 80, SMI 05, SCH 09] for example). Others have sought instead to characterize and interpret the deviations from this regularity [DRE 04, SAN 12]:

– a primary distribution (concave curve, Figure 3.8(b)) describes a situation where the second largest city is smaller than what would be expected from the law;

– a convex distribution (Figure 3.8(a)) characterizes, on the contrary, a system of cities where the cities following the largest city are of a larger size than expected.

The first case corresponds to systems where a few forces are involved and where the largest city polarizes the system. The second describes a system with little integration. The adequacy to Zipf's law is however interpreted as a situation where the forces are numerous and are acting in many varied directions, reflecting a system whose entities are well-integrated in the system.

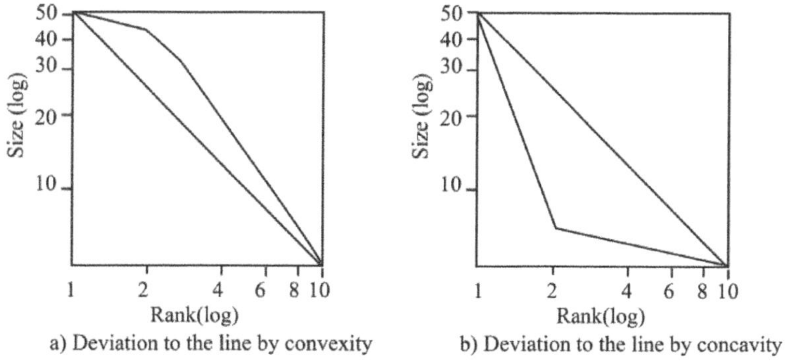

Figure 3.8. *Theoretical shape of the curve crossing cities' rank in the abscissa, and the number of inhabitants in ordinate (source [GRI 11])*

The shape of the rank-size distribution can therefore be used to characterize and interpret the change of the hierarchical organization of a settlement system between different dates. Figure 3.9 thus illustrates the persistence of a regular distribution for the case of South-African cities with more than 5,000 inhabitants during the 20th Century [GIR 09]. On the contrary, Griffin is interested in the oscillation over time between convex and concave distributions of the urban centers of the Tiwanaku valley between 1500 BC and AD 1000 [GRI 11]. The concavity period corresponds to the emergence of an urban system well integrated around a main pole, whereas the return to a convex distribution marks the decline of Tiwinaku that was reflected by a dispersion of the population.

The comparisons in time require, however, a certain degree of caution, the settlements considered being able to change delimitation and meaning over time. While many works have focused on methodological aspects (such as the choice of which probability laws to use), the first question to pose, faced with change in the rank-size distribution of a system of cities, is of an ontological nature (Chapter 1). Berry and Okulicz-Kozaryn [BER 11b] hence form the hypothesis that the deviation to Zipf's law observed for the hierarchy of American cities would be essentially due to the mobilization of empirical entities not corresponding to what are today the real functioning areas of cities. With respect to the top of the urban hierarchy, the urban areas consisting of several urban units are the areas that make sense and not the urban units themselves. This reflection again illustrates the importance of a critical discussion on what is the reference when studying a system's transformations.

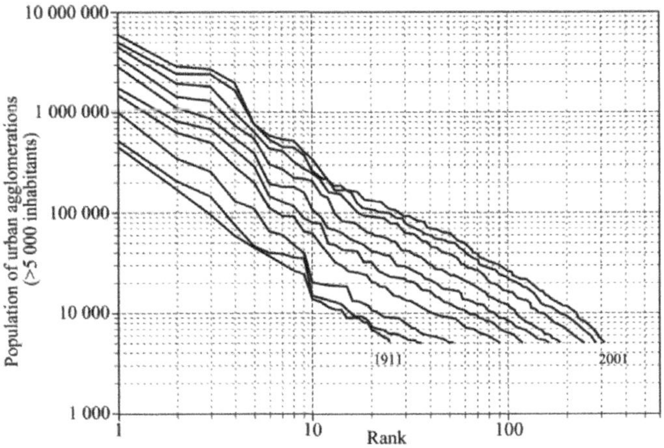

Figure 3.9. *Evolution of the hierarchical structure of the South-African cities with more than 5,000 inhabitants during the 20th Century [GIR 09]*

These four examples present a temporal perspective of indicators conventionally used in a spatial analysis approach. It is a way to account for the three dimensions that are thematical, spatial and temporal. Some of these examples mobilize statistical models, for the sole purpose to represent space at time t, and to account for the evolution by repeating this representation over time.

3.2.2. *Change, temporalities and statistical explanations*

The statistical models can be interpreted as using a mathematical formalism to study the variations of a variable as a function of the variations of other variables said to be explanatory. The simple or multiple linear regressions are examples. In the previous section, regression analysis has been used on two occasions. The first time it allowed us to give a account of a spatial organization, with the spatial regression describing the variations of the population density (log) as a function of the distance to the center (Clark's model). The second served to describe a hierarchical organization, with the regression expressing the variations in population (log) as a function of the rank (log) (model of the rank-size rule). In these two examples, the models, simple linear regressions, are used to summarize the structure with the help of the slope of the regression line.

In the context of a more classical use, that is to say in an "explanatory" model, a statistical model can be constructed (multiple regression for example) at several dates and we can compare the structures highlighted by each of them through the respective roles of the explanatory variables. In this case, the interpretation of change is rather qualitative, even if, as in the previous cases, the trajectories of parameters whose value and significance are evolving over time can be built.

Statistical models can also be used to "explain"[7] change more directly. In this case, time can be summarized in the variable "to explain" which is then an evaluation of the change of the phenomenon of interest (entities' population variation between two dates, change in land use, for example). This type of model makes it possible to formalize the generic framework proposed by Durand-Dastes [DUR 90] to describe and explain the differentiated evolution of places (for example a set of municipalities). He proposed three categories of variables referring each one to an explanatory domain:

1) The ecological variables that refer to the economic, social, and physical profile of these "municipalities" taken as the statistical individuals.

2) The historical variables that refer to both the past trajectory and time-stamped events.

3) The spatial variables that characterize the relative position of the municipalities, their accessibility, their neighborhood, etc.

Furthermore, in a predictive context, the question underlying the construction of an "explanatory" model of change can be derived with regard to the three components: where will change take place, how much change will occur and when will this change take place? On the other hand, this type of model raises the question of the capacity of a statistical model to integrate the spatial variability of the causes of change.

7 We have chosen to put the term "explanation" in between quotation marks when it refers to statistical terminology. In Chapter 4, we will discuss the different meanings of the explanation with a spatial analysis approach.

The example of Serneels and Lambin [SER 01] is interesting in this sense. They question the ability of a statistical and spatial model to help and understand the forces of change of land-use in the region of the national reserve of Maasai Mara in Kenya. This region is experiencing a strong competition between different land-uses. It covers an area from the lands in the North of the region, called "high potential lands", widely open to agriculture, to the reserve covered with prairies and subject to strict conservation rules. These two areas are separated by an area that acts as a buffer and reserve pasture for migratory species and Maasai pastoralists. Due to changes in land administration in this area in the 1970s, a number of major transformations of land-use have succeeded: fragmentation of parcels and large-scale cultivation of wheat. Between 1975 and 1995 more than 8% of the vegetation of the surface area was lost, for the benefit of agriculture by the multiplication of the number of small farmers-owners and the simultaneous arrival of mechanized agriculture. The authors call up satellite images to identify the types of change leading to the decline of the vegetation cover. Three types of change are identified:

– the conversion to a mechanized agriculture on a large-scale;

– the conversion to a subsistence agriculture when permanent colonies arrived;

– an impoverishment of the vegetation cover.

The authors have built several multiple logistic regressions to explain change associated with each of these types of change for the two periods 1975–1985 and 1985–1995 [WRI 02]. The considered spatial entities are the pixels of 100 m. The authors integrate a formalization of space that we call "active space", by forming the hypothesis of the role of spatial variables such as the distance to the capital of the region, the accessibility to roads, to villages, the distance to water, hence integrating the logic of Von Thunen's model [HAG 65]. These variables describing the situation of entities relative to some spatial structuring elements are combined with the variables describing the site: the altitude, the aptitude of the land, the type of agro-climatic area, the type of landowner, the population density and the population variation between the two dates. The combination of all of these variables allows the explanatory factors of change to be approached. Depending on the type of change considered, the explanatory powers of these factors differ: for example, the probability of change toward large scale agriculture is very significantly explained by the whole set of variables. This is not the case of the probability of change toward a small owners' agriculture involving water access and the proximity of the park, which is crucial for their pasture in times of drought.

This approach allows the authors, on the one hand, to identify the most decisive factors to explain the spatial distribution of change according to the type of change, and, on the other hand, to assess whether there has been any stability of the explanatory factors of change over the two periods. The authors conclude on the ability of such models to help better understand change: they remain however inquiring on the determination of the causes and underline the difficulty of distinguishing correlations and causes, as well as determining the meaning of their relationships. One of the main difficulties in the study of land use changes comes from the fact that the driving factors of change have a large geographical variability.

A second example of interest should be presented here: the modeling of the capacity of the entities of a settlement system to persist through time [SAN 97a, FAV 98]. The analyzed system is composed of 680 settlements. With each settlement is associated its date of apparition during a period from the 1st to the 11th Century. Among those, some have lasted until nowadays, while others have been abandoned. The question is to identify the factors which "explain" this difference. The variable "to explain" is a dummy variable: the settlement persisted (1) or not (0). The author explores the effects of the following four factors: the site, situation, hierarchy and accumulation effects over time. The site effect is tested from the indicators describing the soil quality, the altitude and the topography. The situation site effect was measured by the settlement's position in relation to the road networks of the considered period as well as the proximity of other settlements. The hierarchy was measured by variables describing the functional and hierarchical level of the settlement (area, quality of materials, presence of craft, of religious building, etc.). Finally, the accumulation includes indicators of the historical context as to whether or not there was a previous occupation [FAV 98, FAV 12]. The model implemented is a logistic regression. It has allowed us to show the discriminating roles on its persistence of the hierarchical level of the settlement and of its situation at a road intersection. Everything being equal regarding these two factors, the differences in site or of historical context or even of proximity to a larger settlement do not differentiate the capacity of the settlement to persist through time.

In each of these two examples, the variable to be explained (the dependant variable) describes a change. It concerns in both cases a change in status (land-use or existence/disappearance of a settlement). The approach would have been the same in the context of a quantitative variation (variation of the share of a category of active population at the level of municipalities or variation of a pollution indicator at the level of river sections for example).

Other formalizations of variables that are to be explained using the same type of models, make the temporal dimension directly explicit. This is the case, for example, of a model where the variable to be explained is the opening date of a brand of supermarket at the city level [LAU 86]. The explanatory variables called up are the size of the city, the distance to the parent company, the political color of the city and the existence of another supermarket of the same category. In this model, the opening of a supermarket is interpreted as an "innovation", and the model formalizes a diffusion process along the urban hierarchy. The application in the case of Sweden between 1927 and 1976 reveals the main role of the size of the city and the absence of role of the distance to the company headquarter. It also shows the existence of an avoidance strategy for the supermarkets coming from a cooperative chain when a competition strategy predominates for the others.

The approach which consists of modeling change, eventually allows the trends to be extended by reasoning "everything else being equal" (*ceteris paribus*) by identifying, the average law describing the change depending on the overall structure of associations between variables. Two points do however warrant attention:

– For some phenomena, explanatory factors vary over space. To remedy this problem, some authors have developed specific statistics models. The "geographically weighted regression" [GWR, FOR 02] is thus a regression method that allows us to take into account the localizations of the entities and their local spatial dependency. In such a model, the regression parameters vary in space, which can improve the quality of the model.

– The model enables us to identify *where* the change is more likely to happen but it does not allow us to identify *when* this change will take place. These approaches are therefore instead related to descriptive approaches, which allow us to progress with the understanding of the spatial organization of change.

3.3. Understanding the evolution of a spatial system's entities

The question of the follow-up of spatial entities generally arises when the latter make sense, and the analysis will generate a reflection on the dynamics of the system that they compose. In a first example, time is formalized through a given variable for a series of dates. The concern is about representing the demographic and economic trajectories of cities and the evolution of each city is considered independently from those of the others. The objective is to characterize the diversity of trajectories from a clustering that allows the shape of these trajectories to be categorized. In the second example, the question is to identify the evolution of employment centers in a metropolitan area between two dates, both in terms of existence and spatial boundaries. Spatial interactions enable the changes to be identified. In this example, the method consists of building *a priori* types of change and performing a classification of employment centers following the rules hence defined. The third example is about modeling the land cover change in a perspective of planning urban sprawl. Two models are coupled: a model where the status of an entity depends on its previous state and a model where it also depends on the states of neighboring entities.

The point of interest of these three examples is to simultaneously illustrate three different types of objective and three adapted methods: in the first example, the objective is to *describe* the types of evolution; in the second, the concern is to *build* the types of evolution and in the third, it is about *reproducing* these evolutions. In all three examples, time is envisaged as a sequence of states ("time unfolded"). In contrast, space does not play the same role: it is a "support" in the first example and "active" in the two following, although in two different ways.

3.3.1. *Constructing, describing and categorizing trajectories of evolution*

The first two examples make up a generic case where the challenge consists of identifying *a posteriori* the types of changes: they are based on methods of multidimensional data analysis that are appropriate to account for the diversity of the developments of a set of entities, namely the clustering analysis and correspondence analysis. The statistical individuals are clearly identified and described by the measure of the same phenomenon at several dates. In this example, the concern is about cities, described first by their populations at different dates (univariate case), and then by the evolution over time of the structure of the active population in terms of economic activity categories (multivariate case).

3.3.1.1. *Evolution of the system of European cities between 1600 and 1990*

To analyze the evolution of a system of cities described by a single variable, the methods of clustering are appropriate: in the associated statistical table, a row represents the trajectory of the "city-entity" relative to the targeted variable. The example that we are proposing concerns 450 European cities described by their population at 9 dates (1600, 1700, 1800, 1850, 1950, 1960, 1970, 1980 and 1990) [BRE 99]. It concerns an urban system with a strong growth: between 1660 and 1990, the population of the cities has been multiplied by 30 (it increased from approximately 10 million in 1600 to 300 million in 1990). The purpose of this longitudinal analysis is to follow up the relative position of each city in the urban system, and to describe how the system growth is distributed between the cities. The method used is a *hierarchical ascending cluster analysis* using the chi2 distance. The latter is appropriate for the analysis of contingency tables, because it ensures the distributional equivalence, that is to say that the similarities between the entities will be calculated from the distribution profiles, the weight associated with each category being inversely proportional to its weight in the statistical population [BEN 80, SAN 89, LEB 06]. In this example, this means that the similarities between the cities' trajectories will be calculated from the evolution profiles, and that these are approached relative to the evolution of the urban system taken as a whole.

The results of the clustering (Figure 3.10) highlight a first distinction between two major groups of cities: those that are in relative growth and those which are in relative decline (i.e. either in absolute decline or in a lower growth than the whole system). This first major differentiation supports the hypotheses of a selection process between the cities leading to the hierarchization of the urban system [BRE 99]. This process benefits the cities at the top of the hierarchy, while most of the cities that are smaller in size benefit less from the growth. Each of these two major groups can be further subdivided, highlighting different time frames over the four centuries. For example, regarding the class of cities with a relative growth, the structure of the urban system is divided between the cities that have benefited from the 19th Century growth (the large cities of the old industrial regions such as Belgium, Germany, Holland), the cities whose growth has been continuous (France, Germany and Italy) and those whose growth is recent (Mediterranean arc and Eastern countries). The cities that are found at the interstices of this framework are generally smaller at the end of the period, and have least benefited from the growth over the whole period. The choice has been made to distinguish nine types of trajectory. This division reveals that the relative growth or decline appear with different intensities and temporalities. As an example, among the cities with low relative decline (Figure 3.10), there are three types of trajectory:

– cities that have evolved the same way as the whole system;

– cities that have first lagged behind compared to the whole system's growth and that have then been catching up with a growth stronger than the rest of the system;

– cities of which the start has been fast, but that have then seen some kind of breathlessness compared to other cities.

This interpretation goes in the direction of the hypotheses expressed by the authors, namely of a differentiated growth of cities due to a selection process. The *a posteriori* mapping of the types of trajectory complements the graphics of average trajectories of the classes with regard to the organization of potential interactions leading to the hierarchization of the system.

3.3.1.2. *Evolution of economical specialization of American cities between 1970 and 2000*

The second example appears as a complement to the first (methodologically and thematically) by an analysis of the manner in which the cities adapt to economic innovations: it concerns a longitudinal analysis on the economic specializations of 737 cities[8] of more than two million people in the United States described by the distribution of the active population in 26 sectors of activities over four dates between 1970 and 2000 [PAU 10]. The data is thus a typical case of the

8 Here, the city is considered with the meaning of the definition of "metropolitan area".

spatio-temporal cube presented by Berry crossing space (the cities), theme (the categories of activities) and time. The correspondence analysis allows us to summarize the respective similarities between rows and columns of the contingency tables. It is used here to give account of the successive states of the system and validate what the authors call the cities' coevolution, referring to the manner in which the cities interact through a combination of competition and hierarchical diffusion of innovations.

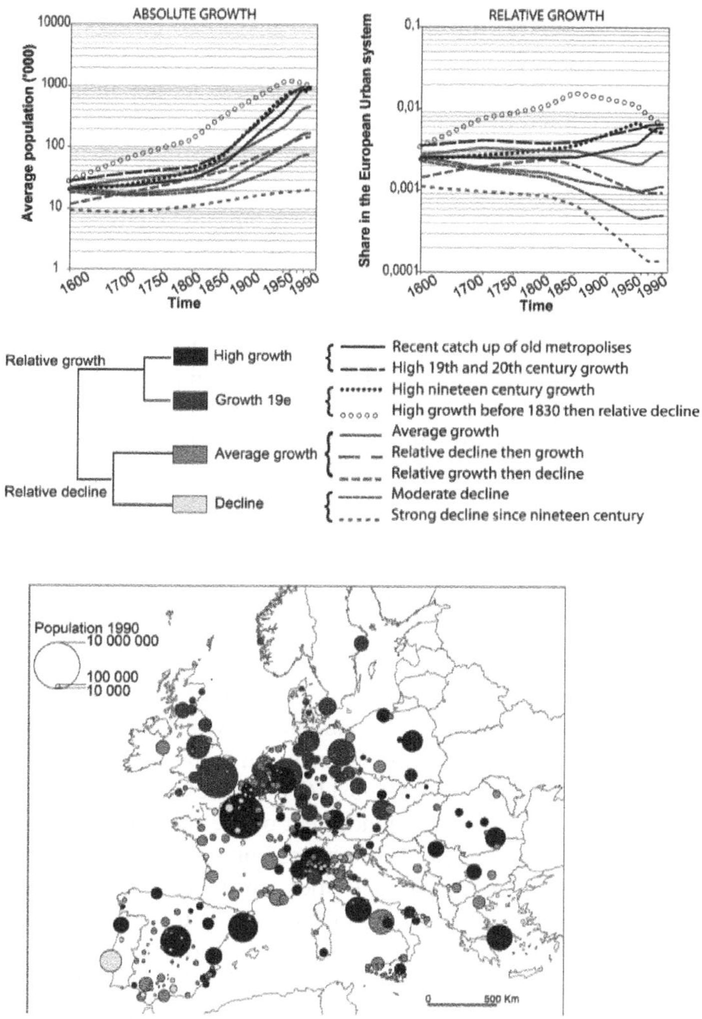

Figure 3.10. *Classification of the European cities' evolution trajectories between 1600 and 2000 (source: [BRE 99] revised L.Lizzi)*

To apply a classic multivariate analysis, it is necessary to refer to a two-dimensional statistical table. The choice must then be made to project time (dates) on the variables or on the individuals. This choice has implications. In the first case, the profile of an statistical individual, here a city, will be the distribution of its active population counted at each of the four dates in the 26 economic activity categories (4 × 26 columns). The "average profile" corresponds to such a distribution on the whole set of cities. The choice may be to project, on the contrary, time on cities. In this case, each city will appear as many times as there are dates, that is four times in this example. Figure 3.11(a) illustrates each of these two choices. If the aim is to "follow" the categories of activity over time in the way that they differentiate the cities, then time has to be projected on the variables. As the authors are focusing here on the differentiations of the trajectories of cities, the most appropriate choice is to project time on the cities (the individuals). The correspondence analysis will therefore be carried out from a table with 4 × 737 rows and 26 columns. A time-stamped-city is a point in a space with 26 dimensions, and a city is characterized by the trajectory linking the four time-stamped-individuals associated. The analysis' referential, if all the time-stamped-cities are "active" in the analysis, will be an average city with all the dates mixed-up. It is also possible to choose a different referent, for example the first date of the period, in order to highlight the change in the cities' differentiation structure relative to this date in particular. Only the pieces of information relating to the cities at this date will then have to be considered as "active" individuals, the others being considered as "supplementary" individuals (they will be represented but will not participate in the identification of the differentiation structure).

In the example, the authors make the choice to read the cities' coevolution relative to an average distribution structure of the activities over the whole period. The first two factors resulting from the analysis summarize these differentiations (25%). The trajectories are constructed *a posteriori* on the factorial map crossing the two first factors (Figure 3.11). This process, if it applies to a relatively small number of entities, is interesting because it simultaneously allows us to highlight the evolution of the whole system and identify the specific characteristics. The trajectory with the dotted line in Figure 3.11(b) represents the average evolution of cities of this urban system, with a displacement of the specialization from industry activities (manufacturing) toward the services, on the one hand, and of "old" activities toward the high technology activities, on the other hand. The position of the trajectory of each city, its length, its breaks and its different orientations, bring forward both the diversity of changes on the economic plan and the diversity of rhythms and steps. By crossing these trajectories with the cities' rank in the urban hierarchy, the authors were able to bring forward the diffusion of innovation cycles along the urban hierarchy.

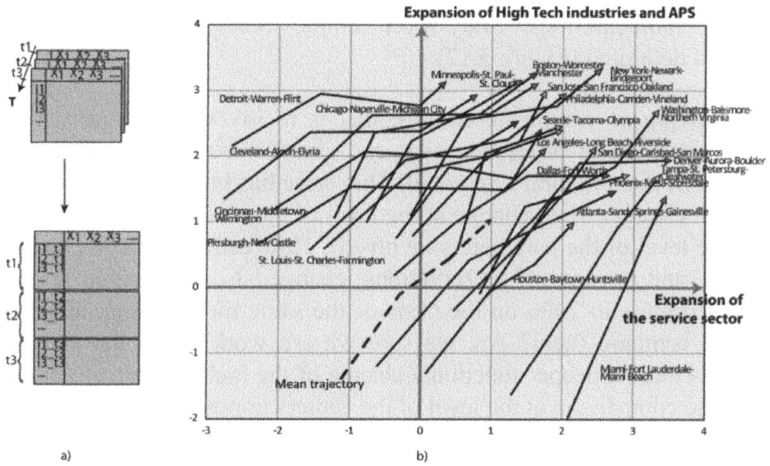

Figure 3.11. *Spatio-temporal analysis of cities' economic specialization in US [PAU 10]*

This type of analysis is extremely rich in revealing the evolution structures from a multivariate description of the entities. It exploits the fact that the temporal dimension is explicit (time "unfolded"). The evolution map can then be represented ("support" space). They can be conducted on the contingency tables or heterogeneous tables referring to a diversity of thematics, such as, for example, a set of basins described by flow measurements over time.

3.3.2. *Identifying a priori defined types of change to classify evolutions*

The previous example was based upon easily formalized evolutions in a statistical table: the objects are well identified and measurements easily repeatable in time. In the case that follows the situation is a little different. It concerns assessing the change of a set of composite objects, the construction of which is the result of a multi-criteria analysis with several steps. This is the case of the employment centers of the urban area of Paris. They have been defined on the basis of statistical data at the level of the municipality from the year 2000, with a method combining a multi-criteria approach to identify the elementary entities (municipalities) that can be considered as the core of the centers and a spatial analysis to then aggregate them into employment centers [BER 11a]. The multi-criteria approach has called up indicators describing, on the one hand, the concentration of jobs and, on the other hand, the attraction exerted by the municipalities on the rest of the territory. The spatial analysis method has called up indicators to identify the shape of interactions between the municipalities. Sixty-seven employment centers have thus been identified and delimited (Figure 3.12).

The follow-up of composite entities, that is to say whose construction is the result of a sequence of processes, as is the case here, raises the problem of the identification of the temporal referential. This issue has been raised in section 3.1.3: how to detect change, when change at the level of the observations may also induce change at the level of the parameters involved? How can we interpret change when observations and means of observations change. Is it possible to construct employment centers in 2006 on the basis of the same methodology as that used in 1999 and to compare them? Are we sure we are working on the change of the objects? Or is there a deeper functional change of the metropolitan system involved that makes the comparison at the level of the centers impossible?

Faced with the difficulty of resolving such an issue, the choice was made to use a classification method based on decision tree logic. Six types of theoretical change have thus been identified *a priori*:

1) disappearence of an employment center;

2) appearance of an employment center;

3) spatial extension of a previously existing employment center;

4) decrease of the spatial extent of a previously existing employment center;

5) splitting of a center into two new employment centers;

6) fusion of two employment centers.

The type of categories of change presented in Chapter 2 in the case of simple objects is encountered once more. Since composite objects are dealt with here, the change must be identified with criteria similar to those that have served to construct the objects. The point is then to develop the criteria that will allow us to assess the most probable change. To identify the cases of “appearance”, a score has been calculated for all the municipalities that did not belong to a center in 1999, on their ability to attract. Then, for each center existing at the first date, four scores have been calculated to assess: (1) its capability to polarize; (2) its propensity to spread over the neighborhood; (3) the intensity of its internal cohesion; (4) its propensity to interact with other neighboring centers. A decision tree was then built by establishing for each type of change, the criterion or criteria to analyze, and the associated thresholds. The score on the capability to polarize has for example been called up on all the municipalities to identify a possible “appearance” (high score) or on all the centers to identify, on the contrary, a “disappearance” (low score). The score measuring the strength of the internal cohesion has, for its part, been used to identify the disappearance, the retraction, or the scission of a center existing at the first date. We will not present in detail here the whole tree and the associated thresholds. The objective of this example was mainly to illustrate the difficulty in tracking a composite object resulting from a process of construction established at time t. The process itself mobilizing evaluations of interactions between the places makes it difficult to construct a concrete and stable referential in time. In this case, the classification procedure allows a supervised procedure to be put in place, to identify change *a priori*, and not to assess it *a posteriori* on the basis of the comparison of two states. Figure 3.12 illustrates the few transformations that have been identified: overall, the main tendency is toward stability. However, a few extensions should be noted: they concern the most dynamic centers, which extend in the direction of the satellite towns [BER 11a].

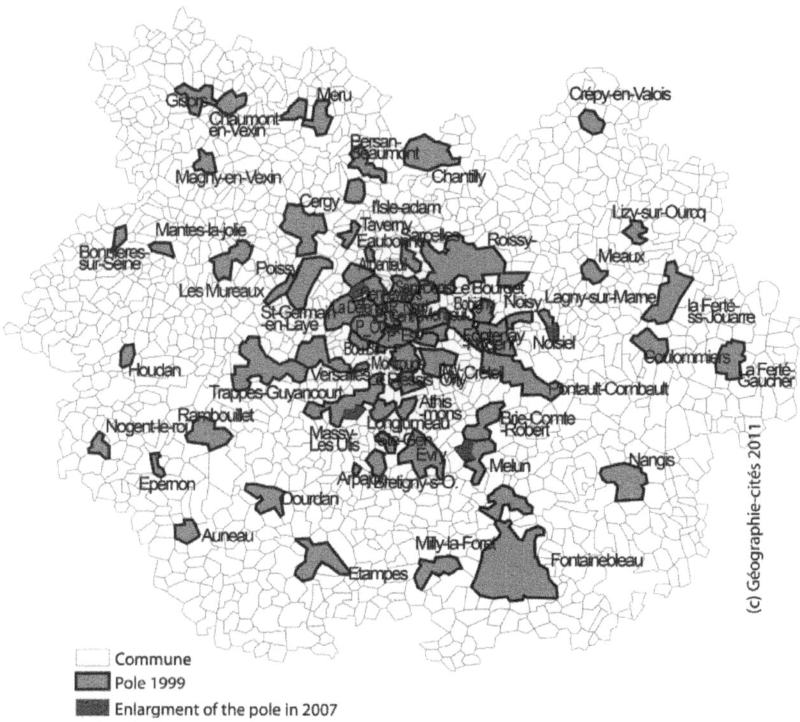

Figure 3.12. *Identifying composite objects' changes: employment centers in the Parisian urban area (source: [BER 11a])*

3.3.3. "*Unfolded*" *time and forecast*

The analysis of observations made on regularly spaced dates (for example, the population of cities each year) refers in statistics to a domain named "time series" or "chronological series analysis". A time series is denoted by Yt, {with t in Dt}. The analysis of time series is a classical field that has given rise to numerous methods seeking to decompose the series to identify its different components and make forecasts. They organize themselves into families whose genealogy dates back to the autoregressive (AR) models (models where Yt is a linear function of Yt-1 and Yt-2) and to the mobile averages over time (MA) [GOU 95]. The models are differentiated according to whether they process states with continuous values (family of models based on AR-MA logics), or they deal with discrete states (Markov chains). Without giving much detail about the different families, it is interesting to note the difference in perspective that these models propose: time is "unfolded" and oriented there; a state at time t is explained by the prior states. But in this "unfolded time" perspective, a date (t) has a specific status compared to earlier dates. It thus differs from the "snapshot" approach where each date t has the same status and is interpreted in its succession. It is to be noted that these families of models have also integrated the spatial dimension: a state at time t may be explained by the prior states, and most often for geographical issues, the prior states of the entities in the neighborhood.

To finish this panorama, we develop the example of a model based on Markov chains, a model where the states space is discrete. This method is particularly suitable to model the process of change in land cover, and therefore complements the examples presented above. The example that serves as a support here refers to a prospective analysis framework of the urban sprawl around the city of Belfort (east of France) developed by Antoni [ANT 06]. The data on which the model is based are of *field* type: the land cover in 13 categories at three dates 1955, 1975 and 1995, around the city of Belfort. They have been explicitly produced to be comparable. They have allowed calculating that the city has spread-out by 62% over the period, growth that is rather concentrated on the second period.

The principle of a Markov chain is based on the property that the future evolution solely depends on the current value (the present), with the hypothesis that the effect of the past on the future is summarized in this present state. In such a process, the probability that a cell is in a certain state at t+1 therefore depends on its status at t (Markov chain). To define a Markov process, the whole set of transition probabilities between two states must be assessed. It refers to the conditional probabilities $p_{k,k'}$ to move from the state k to the state k':

$$p_{k,k'} = P(E_{i,t+1}=k/E_{i,t}=k')$$

where $E_{i,t+1}$ is the state of the cell i at time t+1, and k and k′ are two modalities among the set of possible modalities (that is for this example, the 13 categories of land occupation).

To "calibrate" the Markov chain, namely to calculate the probabilities of transition between two states that allow the observed changes to be reproduced, the author uses the transition matrices observed at the aggregate level, that is to say, the matrices crossing the land cover at the dates t and t+1: at the intersection of the modality k at t and the modality k′ at t+1 is the number of pixels[9] whose land cover has shifted from the modality k to the modality k′: this quantity is denoted by $\Delta N_{kk'}(t)$.

The ratio of this amount of area having recorded a transition from the state k to the state k′ ($\Delta N_{kk'}(t)$) to the number of pixels whose land cover is k at t (noted N_{kt}), expresses the share of the area occupied by the modality k that has experienced a change of land cover in favor to the modality k′. This ratio can be assimilated to the probability of transition from a state k toward a state k′, noted $p_{k,k'}$:

$$p_{k,k'} = \Delta N_{kk'}(t) / Nkt \text{ with } \Sigma_{k,k'} p_{k,k'} = 1$$

In the example, the transition matrices associated with the two periods are quite different because these periods correspond to different phases of the urban sprawl and have very contextual values, such as the expansion of the collective built area from the 1960s, or the construction of highways. The choice was then made to smooth these values by calculating the transition probabilities from a weighted average of the probabilities calculated from the two matrices. This is the same as taking into account the two states previously observed. Figure 3.13 shows the results of the shares of land cover simulated for 2015 from several weighting systems associated with the two periods. Such an approach consists of varying the weighting system and adopting the situation corresponding to the best calibration.

As it is used here, the implementation of the Markov chain is deterministic since its objective is to extend a tendency, hypothetically in the event where change would obey the same mechanisms as between 1955 and 1995. With regards to models of the land cover change, the Markov chains are used to predict the amount of change at the global level but also to simulate the localization of these changes. In the example, the Markov chain is coupled with a model of potential to locate the changes taking into account the states of the neighboring pixels. Thus, once the number n of pixels that will change status is known, the potential model allows us to calculate for each pixel a score of "change potential" in function of its state and the states of the neighboring pixels: the potential state of a pixel is influenced by the neighboring states inversely to their distance. A weighting system formalizes this influence according to the rules, specific to each state: for example, a built-up pixel will have a strong influence on a non-built-up pixel in its neighborhood, whereas a forest pixel will have a low influence on the change of a built-up pixel. The space is

9 This can also be measured by the area of polygons.

made “active”, and the model integrates the spatial dependency. Other simulation methods based on taking into account the effect of the neighborhood will be the subject of Chapter 4. It seemed important to us to introduce, in addition to the two previous examples, these families of models which, with the aim of forecasting (or prospecting), introduce a different status to the temporal dimension.

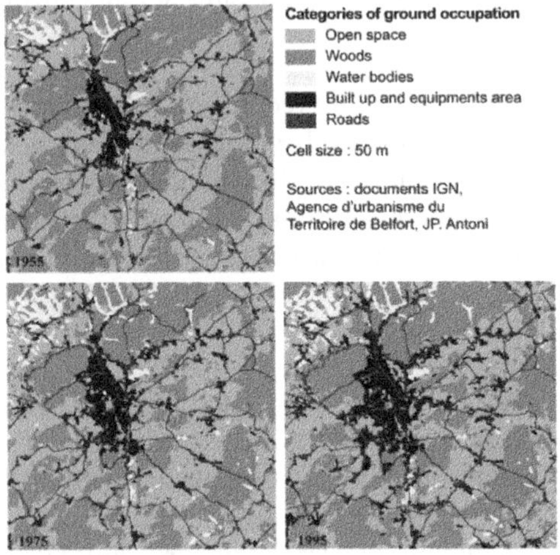

Observed Values (%)

land use categories / Years	*Open spaces*	*Forest*	*Water bodies*	*Single family home*	*dense built-up area*	*Group dwellings*	*Amenities*	*Framing struct.*	*Industrial and com. Area*	*County roads*	*State roads*	*inter-change roads*	*stations*
1955	46.87	38.44	2.29	2.59	2.86	0.13	0.33	0.17	1.06	4.04	1.21	0.00	0.01
1975	42.57	39.08	4.18	3.07	3.28	0.24	0.34	0.24	1.31	4.40	1.25	0.05	0.01
1995	36.88	41.54	4.09	4.56	3.83	0.29	0.60	0.34	1.94	4.43	1.33	0.16	0.00

Simulated values for 2015 (%)

Transition matrix and weighting system used	*Open spaces*	*Forest*	*Water bodies*	*Single family home*	*dense built-up area*	*Group dewellings*	*Amenities*	*Framing struct.*	*Industrial and com. Area*	*County roads*	*State roads*	*inter-change roads*	*stations*
(1) 'Mat. 1955-1975	34.52	41.88	5.45	4.32	4.12	0.36	0.55	0.37	2.02	4.83	1.35	0.22	0.00
(2) 'Mat. 1975-1995	32.33	43.24	3.99	5.83	4.34	0.34	0.84	0.43	2.52	4.48	1.38	0.26	0.00
(3) Mat. 1955-1975 (0.5) ; 1975-1995 (0.5)	33.42	42.56	4.72	5.08	4.23	0.35	0.70	0.40	2.27	4.65	1.37	0.24	0.00
(4) Mat. 1955-1975 (0.75) ; 1975-1995 (0.25)	32.77	42.97	4.28	5.53	4.30	0.34	0.78	0.42	2.42	4.55	1.38	0.25	0.00
Average of the 4 previous	33.26	42.66	4.61	5.19	4.25	0.35	0.72	0.41	2.31	4.63	1.37	0.24	0.00

Author : JP. Antoni ; Logiciel : CWS (Prog. G. Vuidel)

Figure 3.13. *Evolution of the land cover around Belfort and simulation results [ANT 06] (revised L.Lizzi)*

3.4. Conclusion

The examples presented in this chapter illustrate the diversity of insights that can be called upon to analyze change. We have proposed a grid of presentation based, on the one hand, on the geographical level at which the question of the change is raised, and, on the other hand, on the type of formalization of space and time. Table 3.1 summarizes the diversity of approaches and of combinations of approaches called upon to answer each of the thematic questions. These approaches cover the three categories of methods presented at the beginning of the chapter, namely analysis of geographical data, spatial statistics and spatial modeling. According to the objectives, it leads to the analysis of the evolution of an organization (section 3.2.1), to the identification of the factors of change (section 3.2.2), to a representation of the evolutions of spatial entities (section 3.3.1), or, finally, to the highlighting of the change process itself (sections 3.3.2 and 3.3.3).

Section	Example	Follow-up level	Formalization of space	Formalization of time	Method
3.2.1	– archaeological settlements – postal roads – public transport user	Global organization	active	"Unfolded"	Spatial statistics (geometric center, ellipse)
3.2.1	– Urban hierarchy	Global organization	support	"Unfolded"	Rank size rule
3.2.1	– Urban spatial extension	Global organization	active	"Unfolded"	Gravity model Clark's law
3.2.2	– Change of land use	Global organization	active	"Folded"	Logistic regression
3.2.2	– Settlements' durability	Global organization	active	"Folded"	Logistic regression
3.3.1	– Population dynamics of an urban system	Entity	support	"unfolded"	Clustering analysis
3.3.1	– American cities' specialization	Entity	support	"Unfolded"	Correspondence analysis
3.3.2	– Evolution of the employment centers	Entity	active	"Folded"	Decision tree classification
3.3.3	– Change of land cover	Entity	support	"Unfolded"	Markov chain

Table 3.1. *Summary illustrating the diversity of cases*

In the majority of the cases handled, there are issues concerning the referential, both in terms of processing and interpretation of the spatio-temporal phenomenon at stake: in relation to what can the change be described and interpreted? The issue is different in the case where it concerns following in time the overall organization of a spatial system (first part of Table 3.1) or the entities making up this system (second part of Table 3.1):

– The four cases dealt with in the second part of Table 3.1 reveal a variety of links to the referential. In the first case, which gives rise to two examples, the referential chosen refers to the whole period, and the evolution is interpreted accordingly through the system's average evolution. This approach allows us to focus on the specific characteristics of certain entities (in the two examples in section 3.3.1, the concern is about cities) or groups of entities registering a rhythm of change slower or faster than the general tendency of the system. In the second case, the referent is absolute: in the example of section 3.3.2, it relates to the first date of the considered period, as is often the case in applications (in this example, the employment centers in 1999, regarding their existence and their spatial inscription). The point of focus then specifically concerns the change that has occurred from a situation considered as initial, which has been observed at a given time, and that served as a basis for comparison in the representation of change. In the third case (land cover, section 3.3.3), the referential is the intitial situation / transition matrix couple. Indeed this matrix, at the heart of the method, directly represents the manner in which the process of change operates.

– The five cases relating to the following-up of the overall organization of a spatial system illustrate another issue concerning the referential. In the first case (section 3.2.1), where the purpose is descriptive the analysis does not depend on a referential. A gradual representation of differentials is simply represented and interpreted by the user. In the other two cases referred to in section 3.2.1, the referential is constituted by a statistical law or a spatial analysis model, the rank-size law in one case, the Clark's laws and the gravity model in the other. In the case of explanatory models dealt with in section 3.2.2, there is again no referential, the variable "to be explained" representing a measurement of change.

These different ways of understanding change in a spatial system are naturally very complementary. Although proposing different insights, they are all oriented toward an objective of representation and description. The question of *how* is at the center of the questioning. The term process has only been mentioned in the context of the last example. In order to understand the involved processes, the underlying mechanisms of change have to be identified, and then the reason *why* a change has taken place in such a way, why a system has evolved in such a direction etc. Chapter 4 is dedicated to the modeling of these processes that are essential to give an idea of the systems' future dynamics.

4

Exploring the Underlying Processes of Change: Simulation Models

The objective of this chapter is to present different families of simulation models. The aim is to explore spatial dynamics and identify the functioning and effects of the different mechanisms operating in space. These models are anchored in a systemic approach, where the objects of interest (whether they are households, settlements, plots of land or spatial units) are interconnected with their interactions playing a driving role in change. First, in order to establish a link with the approaches outlined in Chapter 3, we propose to compare simulation models with the statistical approach, focusing on their reciprocity rather than contrasts. We then present microsimulation models that utilise the characteristics of each of these approaches. We then continue with models formalized with cellular automata (CA), and finish with multi-agent systems (MAS), with applications on simple or composite objects, as we introduced them in Chapter 1. These two families of models are related to the complexity sciences and rely notably on the concepts of emergence and self-organization. The respective contributions of these different families of models to research will be put forward and illustrated, with their applications in geography and archeology.

4.1. Computer simulation *versus* statistical approach: different points of view about explanation

As it was described in Chapter 3, most statistical methods are not *specifically* designed to deal with change. This is generally accomplished through the construction of variables that characterize (emergence or disappearance of equipment, for example) or measure (population growth rate, for example) the change recorded for a given phenomenon between different dates. In other cases, variables expressing dates (of adoption of an innovation, for example) or durations (for example life duration of a settlement) are adapted to measure and explore phenomena of diffusion or durability. Conventional statistical methods (notably, multivariate data analysis and regression models) allow us to analyze such variables, to highlight how the differences between the studied objects (as "statistical individuals") are structured and to identify trends. Such approaches can be called "data driven" to the extent that information is generated from the manipulation of empirical data. It corresponds to the situation where the problematics of research could be formalized in terms of statistical individuals and variables.

The researcher may have other perspectives in regards to the empirical reality. Let us consider the example of Schelling's model [SCH 71] which is emblematic in the field of simulation in the social sciences. The empirical information consists of remarking that there is ethnospatial segregation in most American cities. This information is of a qualitative nature, because we simply observe that some neighborhoods have a higher concentration of "black" people and others of "white" people, and that this phenomenon is recurring from one city to another. Schelling's aim is to use the model to reflect on the causes of these concentrations. The space is represented by a grid on which a set of agents from two categories, that we will name here A and B, is laid down. At each time step, each agent chooses to remain at the same location or change cell. The rule is the following: if the proportion of agents of the same type as itself in the neighborhood formed by the eight neighboring cells (Moore's neighborhood) is lower than a given threshold s (Figure 4.1), then the agent "moves out" and settles down in an empty cell. The point of this model is to show that it is sufficient for $s = 1/3$ to cause the simulation to show a segregated spatial organization. Daudé and Langlois [DAU 07] have explored in a systematic manner the effects of different thresholds of tolerance according to the spatial density of the agents on the levels of aggregation that emerge (figure 4.2). The conclusion is that there may be spatial segregation of the population without a particular planning policy, and without individuals having segregationist behaviors (since they accept to be a minority in their neighborhood, as long as this minority is not less than one third). Even if the individuals have preferences for a mixed neighborhood, the outcome is this segregated organization.

Such a model is said to be "concept driven": the researcher has a hypothesis on the role of a certain mechanism that she/he implements in the form of an interaction rule between the agents and their environment. This rule operates for each agent, at each time step, and the modeler observes the simulated situation (that is the spatial organization of agents in the grid according to their A or B character) at the end of n iterations. Such computer simulation models are thus inherently dynamic, the driving force of change residing in the interaction rule leading to the move of the agent, or not. Thus, it refers explicitly to the *behavior* behind the change, which offers a different perspective than the one that simply consists of giving an account of the change.

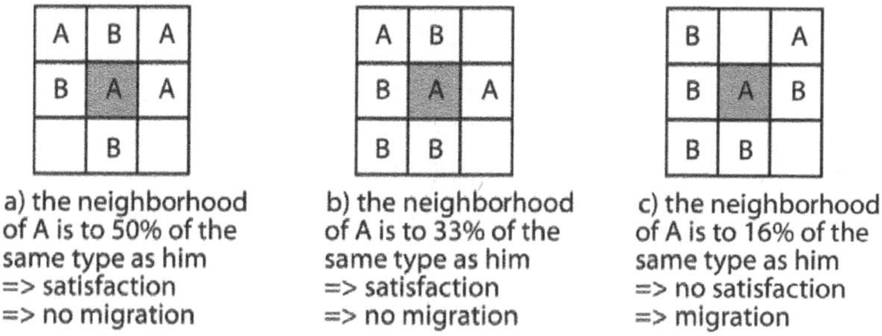

Figure 4.1. *Three different neighborhoods centered on an A type agent (case where s=1/3)*

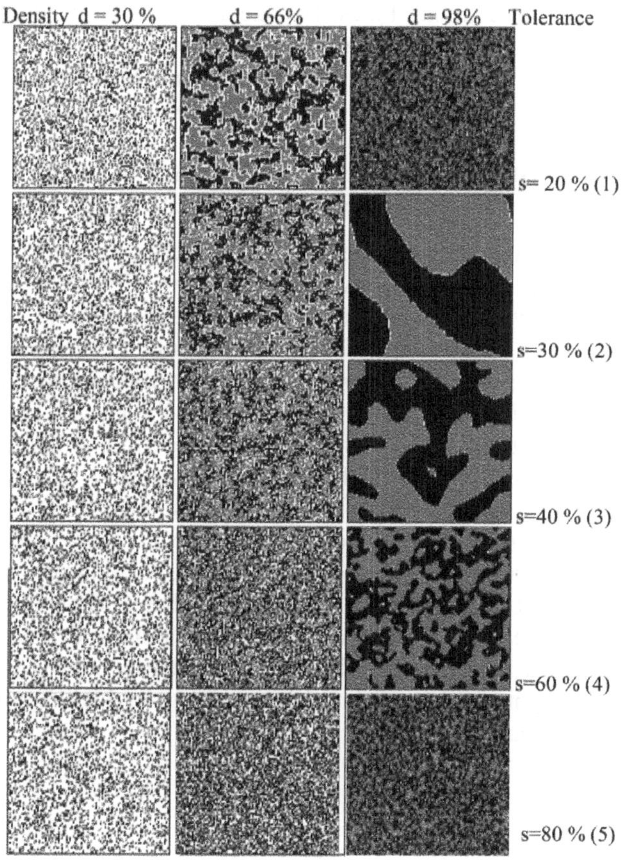

Figure 4.2. *According to the level of tolerance, the simulation resulted in varied levels of spatial segregation (source: [DAU 07])*

Computer simulation has known a growing success among geographers, archaeologists, economists and sociologists for nearly two decades [BAT 05, KOH 00, EPS 96, GIL 99, BEN 04, PHA 07, HEP 12]. Although the success of this form of modeling is increasing, it is more relevant to consider these two approaches (statistics and computer simulation) as complementary rather than competitive. A somewhat systematic analysis of their differences and respective advantages shows this well.

4.1.1. *From covariation to interaction, from differentiation to emergence*

The statistical approach is centered on highlighting similarities between objects (as statistical individuals) according to their characteristics and the *interrelations* existing among their different attributes. The agent approach[1], for its part, focuses on the *interactions*; interactions among agents, on the one hand, and between agents and environment, on the other hand (in the example of Schelling's model, the environment is simply composed of the organization of the other agents in the immediate vicinity). Thus, the key concepts are clearly distinct in the two approaches. In the first case, the issue can be expressed in terms of differentiation and covariation, and in the second case in terms of interaction and emergence.

These differences are visible in the symbolic representation of these methods: a data table comparing statistical individuals with variables in the statistics case (Figure 4.3(a)); following a social metaphor in the case of MAS (Figure 4.3(b)), with agents located inside an environment, having a representation of this environment and acting on it according to their interactions with the other agents [FER 95]. Figure 4.3 illustrates this difference, where the interest lies in the change of the social composition of the high schools of a city or a region. These high schools can be considered as the "individuals" in a statistical approach and as the agents in a MAS. The attributes (public/private status, number of students, number of incoming students, number of outgoing students, etc.), contained columnwise in the statistical table, are strongly associated to the individuals. Considered as properties in the agent approach, they are able to influence the behavior of agents, but they do not appear as central in this formalization and, therefore, are not explicitly represented in the classic diagram proposed by Ferber [FER 95]. This representation highlights the interaction between agents, in this example, by the flow of pupils changing school.

1 In this comparison, we will consider especially, with regard to computer simulation, multi-agent systems (MAS) that we qualify sometimes, in a more general manner, as agent approach, and cellular automata (CA). More precise definitions are given further in the text. For a very thorough presentation of this field, we refer to [PHA 07].

	public/private	Nb of pupils	% privileged	%incoming	%outgoing
School 1					
School 2					
School 3					
...					
School n					

a) Table of statistical data

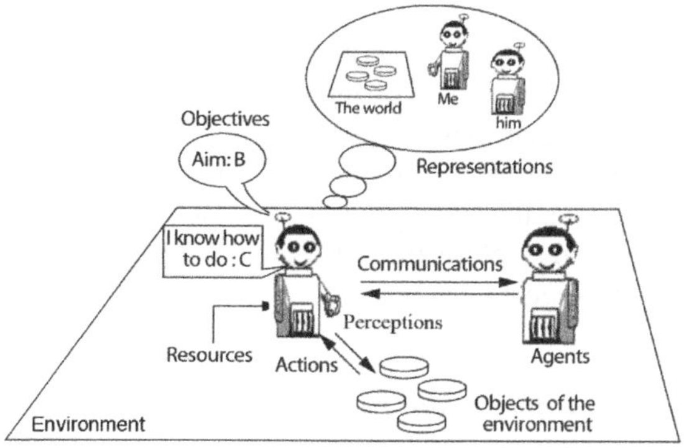

b) Agents in an environment SMA [FER 95]

Figure 4.3. *Formalization of a problematic referring to the differences in social composition of a set of schools following a) a statistical perspective and b) a multi-agent perspective*

The relation to empirical reality differs in the two approaches, which, as we discussed earlier, is linked to differences in the objective of the models developed in a statistical or agent framework. In one case, relatively complete information is available, formalized in the form of a statistical table, and the objective is to develop a model which reproduces the observed facts in the most faithful possible manner (the point is to obtain the best fit between model estimates and observed values). In the other case, a set of rules is established, which allows the production of a stylized fact[2]. In the first case, the aim is to identify and describe the regularities in the *differentiations* between the statistical individuals, and also identify the variables that are discriminant to "explain" (in the statistical sense) these differentiations. In the other case, the objective of the model is to explore the mechanisms at the origin of the emergence of such differentiations. In the first case, the statistical model is calibrated from a set of observations, i.e. the values of the parameters are estimated (the coefficients of a multiple regression, for example) by minimizing the difference between the observed values and the values estimated by the model. The validation step then consists of testing the model on a set of observations that had not been

2 A "stylized fact" is a simplified presentation of a recurring empirical regularity (observable in the form of correlation, spatial configuration or hierarchical organization).

included in the calibration. In the second case, even if the simulation results are compared with the observed data (calibration and evaluation phases), this step is less tightly controlled than in the case of the statistics field. The reasoning is nearer the model of a "candidate for explanation": the set of rules formalized with the MAS enables the reproduction of a given organization (sufficient conditions), but this does not exclude that other sets of rules can also achieve it (conditions not necessarily necessary).

In the statistical approach, the emphasis is therefore placed on the relation between the attributes of the objects of interest, and the constraints attached to the statistical methods imply that these objects are independent [MAC 09]. The agent models, on the contrary, place the interactions in the center of the method. It is because of the interactions that the "objects" are interpreted as "agents". After all, the objective is indeed to explore the forms and properties that emerge from the *interactions* between the agents.

4.1.2. *Different types of explanation: from the percentage of variance explained to the generation of mechanisms*

The points of view of the explanation, which are derived from these two approaches, differ [BUL 05, MAN 07]. The statistical explanation relies on highlighting the covariations of different phenomena. Thus, the term "explanatory" is excessive; the statistical models do not allow us to put forward causal relations. The "percentage of variance explained" by a model refers to the ability of the model to reproduce, from the combination of a set of variables, said share of differentiations between the objects of interest. In the case of high schools in the Parisian region, for example, (see Chapter 2, section 2.3.4), 65% of the interschool differences in terms of grades in the "brevet" (French high school exam) are "explained" by positioning differences on an axis of social composition.[3] However, the statistical model does not allow us to identify the real causes behind the difference in results between schools. As Bulle puts it: "The real mechanisms behave as 'black boxes': the models tend to reproduce the 'inputs' and the 'outputs' of the real sets being studied without revealing their operation secrets" [BUL 05].

The title of one of Epstein's books [EPS 07] about MAS, "Generative Social Science", clearly illustrates the explanation that the author associates with this formalism: to him, to explain a phenomenon means to simulate it ("If you didn't grow it, you didn't explain it" [EPS 99, EPS 07]). If the interaction rules implemented at the agent level generate an organization at a macro level of

3 This axis corresponds to the first factor of a principal component analysis (PCA) performed on the social compositions of the schools.

observation, then these rules are interpreted as "explanatory" of this organization. Other authors emphasize that if such a step is *necessary* to explain a phenomenon, it is *not sufficient* [MAC 09]. On the one hand, different sets of rules operating at a basic level may produce the same configurations at a higher organization level. Furthermore, the objective is also to "understand" these rules. As a matter of fact, Macy and Flache underline that, unlike the models in game theory where causal mechanisms are explicit, it does not suffice to look at the rules to identify the cause behind an event simulated by the model. The modeler is compelled to research the sequence of events that led to the simulated situation. For example, let us suppose that, on applying Schelling's model, and starting with a random initial situation, we observe after *k* iterations, the emergence of two concentration islets of type B agents in the Northeast quarter of the grid (Figure 4.4). To understand the emergence of these concentrations at this location, it is necessary to examine the trajectories of each of the agents that have resulted in one of these two aggregates, i.e. to identify the decisions sequence of each of these agents taken in their successive neighborhoods.

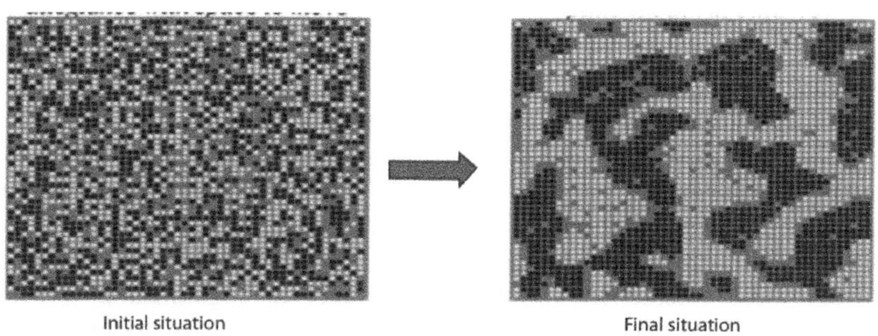

Figure 4.4. *Emergence of a segregated spatial organization from a MAS formalization of Schelling's model (source: [BAT 04]). For a color version of the figure, see www.iste.co.uk/mathian/spatio-temporal.zip*

4.1.3. *Different points of view on multilevel phenomena*

Another area involves different points of view between the statistical and the multi-agent approaches. It concerns the way of taking into account several levels in an analysis. Indeed, many geographical problematics involve several scales of observation and objects falling under different levels of organization. The objective of multilevel approaches is to develop methodologies that take this into account. These approaches are varied in nature. The confrontation of viewpoints associated with statistics and with agent-based simulation illustrates this diversity:

– In statistics, the aim is to make account ("explain" in the statistical sense) of the differentiations that are observed at a given level; for example, the differences in school results at the level of individuals or the differences in health between a set of cities. In the first case, some "explanatory" variables relate to the pupils' level (parents' social class, sex, for example) and others to the level of the school (social profile, private/public sector, number of pupils, for example) which constitutes the context in which the pupils evolve [FRA 14a]. In the second case, on the one hand, there are variables characterizing the cities (size, average incomes, economical dynamics, pollution) and on the other, variables characterizing the region in which the city is located (climatic region, social ambiance, for example) [ASC 13]. The multilevel statistical models enable us to assess the respective effects of these different variables referring to different levels, by reasoning *all things being equal*, on the differentials in terms of academic results or population's health. Such models allow us to draw conclusions on the effect of a pupil's parents' social class on his/her results at the high school "brevet", all things equal with regard to the social profile of the school where he/she performs his/her schooling, and *vice versa*. Thus, a multilevel model allows us to refine the description of the phenomenon of interest.

– In a multi-agent model, the interactions between the entities considered as elementary are formalized (the students in the first example used above and the cities in the second example) and the interest is in the configuration which emerges at a higher organization level. It concerns, in the first case, the manner of concentration of pupils with a certain social profile in some schools, and in the second case, the manner of the spatial configuration of cities according to their populations' health status. The MAS approach is inherently dynamic and multilevel. Indeed, the rules on the interactions between the elementary entities operate at each time step and we can follow the evolution of the configurations that *emerge* over time from these interactions at the higher level of organization through the representation of appropriate indicators. The underlying hypothesis is that the structures observable at a macrogeographical level (spatial segregation in the case of Schelling's model, a rank-size organization in the case of models of the SimPop type, Figure 4.5) emerge from a set of interactions operating at a basic level. The principle is *bottom-up*, the interactions between agents "generating" the structures observed at the higher level, but without the obtained structure being desired or even perceived at the elementary level. The focus is on the plausibility of the configuration which has emerged and which can be characterized from statistical indicators (hierarchical organization, spatial organization, etc.). It is at this higher level that the simulation results will be compared with the observed values, and not at a one-to-one basis as in statistics. Having developed a MAS model with the objective of understanding the redistribution of three categories of Israeli populations (Jewish, Arab Christian and Arab Muslim) in the district of Yaffo between 1955 and 1995 (old Arab city, today a Southern suburb of Tel Aviv), Benenson *et al.*[4] [BEN 02] use this approach to compare the simulated and observed

4 This example is developed in section 4.3.3.

values for two macrolevel indicators, the Moran's index of spatial autocorrelation on the one hand, and the share of Arab population on the other hand. The objective is to evaluate the mechanisms introduced in the model about the residential preferences of individuals. Thus, the calibration of the model is based at the emergent level rather than at the level at which the interactions induce change.

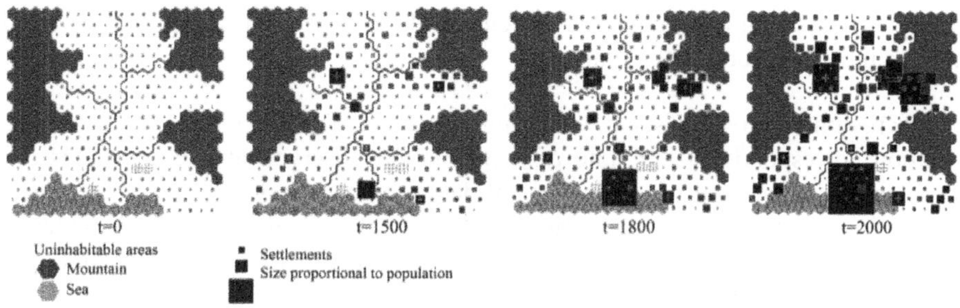

Figure 4.5. *Emergence of a hierarchical system of cities simulated with the SimPop model [SAN 97b]*

While pointing out the differences between the statistical and agent approaches, we are regularly reminded of their complementary nature. Indeed, faced with a given spatio-temporal question, whatever it relates to (an environmental change, the dynamics of past settlements, the social or economic evolution of regions or cities), the statistical models allow us to show trends in the relationships between different phenomena through the analysis of empirical data. Similarly, the computing models allow us to explore the mechanisms that have led to these observed trends. Thus, formalisms such as the CA and the MAS are adapted to explore the hypotheses concerning the effects of interactions between objects, whether they are simple or composite, and the interactions between the considered micro- and macrolevels. The next section is dedicated to microsimulation models. They have an intermediate position relative to the fields of statistics and agent-based modeling.

4.2. Microsimulation models

Microsimulation models are based on the characterization and follow-up in time of the elementary entities of a society, most often individuals, but also households or firms, with the objective of understanding global phenomena. The first models of microsimulation came from economists and sociologists [ORC 57] who advocated to formalize hypotheses about change at the household level in order to better understand the changes in the American economy by major sectors of economic activity: this large gap between the level of modeling (households) and the level of questioning (economic sectors) was then an innovation in social sciences. In the 1980s, the concepts of time-geography developed by Hägerstrand in his advocating lecture "What about people in Regional Science?",[5] are introduced in microsimulation by geographers [HOL 89, HOL 07a, MOE 03] who develop dynamic spatial microsimulation models. These models, centered on the individuals' representations and decisions, allow us to endogenize the origin of changes in the demographic, social and economic structures of a society (the aging of individuals leads to more pensioners, for example).

Microsimulation consists of working on an artificial population. The initial situation corresponds to the state of a population observed at a given time[6] or a construction in the case where individual data are not accessible. In the latter case, an artificial population is constructed by cross-sectioning statistical tables of demographic, social and economic data at the aggregate level of spatial units [CLA 87, WIL 98, STR 05]. The objective is to have a complete representation of the population [CAL 00], and to ensure that this representation is consistent from the point of view of the interdependencies between individuals' spatio-temporal trajectories in terms of family and generational compositions[7] [HOL 07b].

Individuals are described by a set of attributes corresponding to the data provided by a census or a large survey (age, sex, level of education, income, occupation, place of residence and place of work). Transition rules define how these attributes vary over time. The transition from the state t to the state $t + 1$ is formalized in the following way:

At each iteration of the model, each individual is successively examined in order to determine the changes in his/her state. Some changes simply reflect the effects of the passing of time [DAV 01], for example the increase in the age. Others are more complex and relate to either quantitative changes of certain attributes (for example,

5 Lecture delivered at the European Regional Science Association (ERSA) in 1969.

6 Such is the case of the applications of Holm *et al.*'s team [HOL 02] in Sweden, where researchers have had access to a complete database of the Swedish population at the individual level, with a localization provided at the 100 m cell level. This case is rare; complete individual data is seldom accessible.

7 In order to reproduce the process of family formation in a consistent way, empirical work about family planning and the decision to have children were used. Studies about the interactions between the members of a family were also mobilized, notably about the constraints imposed by the choice of one of the members on those of the others (a parent who migrates to change job takes with him, in general, the other members of the family) [HOL 07b].

the income of the individual), or an event (having a child, change of residence or work). Some of these changes are simulated by simple random drawings from a probability table known at the level of the whole of the population (for example, a woman of a particular age, already having *k* children, has such probability of bearing a new child). Others use regression models or logit models to connect, for example, the probability of migration to the individual's age, type of work, gender, income, etc. Such models have been previously calibrated from empirical works at the individual level [REP 04]. Some, less frequently, refer to cognitive processes and formalize individuals' intentions, attitudes and preferences estimated from interviews [HOL 89]. Hence dynamic microsimulation is most often based on probabilities in order to model the behavior of individuals. The transition rules are driven by the need for consistency between the changes in the domains of demography, employment, education, residential market, etc. Thus, a household that emigrates releases an accommodation on the real estate market, a deceased active implies an employment offer for corresponding qualification, a student who completes his studies becomes an employment seeker, etc.

The objective of a microsimulation model is often operational. A coherent artificial population has been constructed and it is evolving in a consistent manner. Therefore, the model can be used to test the impacts of political changes on the society according to different scenarios. This type of model allows us to ask questions of the type "what if?". For example, if a new university is created at such place, how will the students be redistributed? What would be the effects of a change in the immigration policy on the total population in the 22nd Century? What consequences would a change in pension schemes have on the real estate market? In these examples, the question relates to the effects of a change made at the global level (immigration policy, pension plan) on a phenomenon also apprehended at the global level (the total population in 2100, for example). Between the change and its consequences at the global level, the model has simulated the actions of each individual (demographic, choice of residential, of work, etc.). Indeed, the individual level is that at which the true driving forces of change are situated according to the authors of this field [CLA 87, REP 04].

One of the benefits of microsimulation is that the result of the change can be described at any level by simple aggregation of individuals. Regarding spatial outputs, the spatial units, whether they are agglomerations, municipalities, cells, or regions, behave as a receptacle for accounting the resulting individual behaviors. The SVERIGE[8] model [HOL 02, REP 04, HOL 07a] is based on the principles

8 The acronym SVERIGE stands for System for Visualizing Economic and Regional Influences Governing the Environment. Let us note that this term also designates Sweden in Swedish.

outlined above and imposes itself as spatial *and* dynamic. It contains a dozen interrelated modules, respectively dedicated to demographics (including the processes of moving in together, divorcing and moving out of young people), education, employment and migrations. This last module has three phases and a submodel of choices has been developed for each of them:

– the decision of the individual to migrate;

– the choice of a new labor market;

– the choice, within the corresponding region, of a cell of residential localization.

Once the model is elaborated, it can be used to estimate all the phenomena associated with the actions introduced in the model. It has been used notably to explore the consequences of different policies in the area of immigration with regards to the 22nd Century, showing that a difference of 20,000 people in the annual flow of immigration is reflected, within a century, by a decrease of the Swedish population of several millions (Figure 4.6). Alfredsson [ALF 02] focusing on the links between the behavior of households in terms of heating and transportation means, used this same model to show that even scenarios introducing significant changes of behavior at the individuals level, result in a low impact in terms of CO_2 emissions, due to the importance of the forces of inertia. In this last example, the consequences in which the author is interested in, belong to the scope of the global level, but the changes whose effects are to be studied, unlike the previous examples, belong to the scope of the individual level.

Similar principles have been mobilized in the MICDYN model which has been applied to explore different scenarios on the distribution of the population in the Languedoc region. The model was used to better understand the process leading either to an urban concentration or to an urban sprawl around the largest cities, or to a diffusion of urban growth on the whole of the territory [ASC 00, HOL 07b]. The originality of this model lies in the fact that while the dynamics is based on individual choices, the interactions between different spatial dimensions have been taken into account. In particular, entities referring to the mesogeographical level have been introduced. These are the municipalities, which appear as interacting with the individual choices: on the one hand, an evaluation of the social composition of the municipality is made to construct an indicator of attractiveness differentiated according to the social category of the individual concerned; on the other hand, a strong population growth of a municipality is accompanied by a growth in induced employment related to services. Finally, the growth is distributed over the neighboring municipalities when the density exceeds a critical threshold.

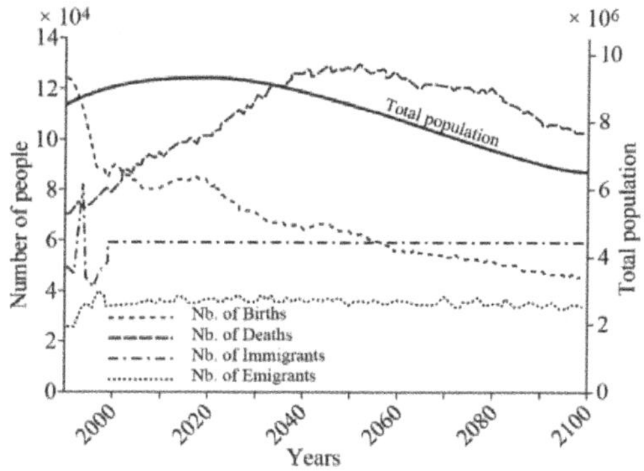

a) Follow up of the initial situation with 60 000 immigrants per year

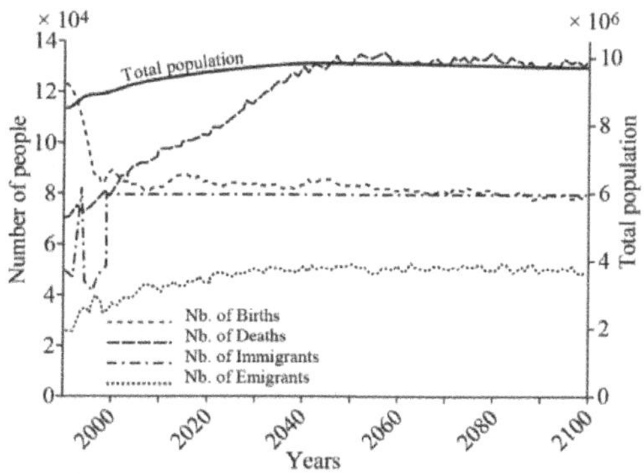

b) Change of politics with a threshold of up to 80 000 immigrants

Figure 4.6. *Simulation with the SVERIGE microsimulation model of the evolution of the population during the 21st Century for two different scenarios regarding immigration policy*

The UrbanSim model is a model dedicated to urban simulation. It has been developed and enhanced during the last fifteen years ([WAD 02]), with a clear operational objective. The model is used as a tool in decision support for testing the effects of urban policies in transport (individual and collective), spatial planning and environmental policies. This model is initially based on a coupling between a microsimulation model and a geographical information system (GIS) that allows to manage and represent the different layers of information at stake (environment, transport, land use) at the same time. The model has been applied to many cities and several other methodological couplings have been progressively introduced. The microsimulation model manages the individual choices in terms of demography (giving birth for example) and jobs. Discrete choice statistical models are used to estimate the localizations of households and jobs. A CA model is used to predict the changes in land use, whose principles are based on neighborhood effects. A MAS is used to model the interactions between the developers, the policies, the jobs and the households. The model has been applied to numerous American cities (Seattle, Detroit and San Francisco, for example) and in the world (Rome, Beijing and Seoul, among others). From an epistemological point of view, this model is the opposite of the Schelling's model mentioned in the introduction of this chapter. It contains a multitude of processes and once well calibrated, it is operational, allowing relatively safe estimates in the short term. Nonetheless, it is not appropriate to explore how interactions operate between different urban processes and identify changes in the founding mechanisms of urban dynamics.

4.3. Computing models: simulation and emergence

As it has been pointed out in section 4.1, the AC and MAS models favor an approach in terms of "generating mechanisms" and place the emphasis on the organization, which emerges at a higher observation level, the driving force being the interactions operating at a lower level. A simulation model can play the role of a "laboratory" where the possible consequences of hypotheses concerned with the formation of the interactions between the considered entities (for example, forming an artificial society, an artificial system of population or an artificial school space) are tested. As such, the idea is similar to the microsimulation models presented above. The difference is probably more attributable to the practices than to the methodologies [BOM 04]. The microsimulation models aim to rebuild, in the finest possible manner, an artificial population in order to test the effects of different types of changes in political or planning areas. More often interested in stylized facts, many MAS applications are relatively parsimonious in terms of modeled mechanisms (such is the case of Schelling's model mentioned in section 4.1). Others incorporate several dimensions but the behaviors of the agents tend to be governed by cognitive principles rather than by statistics. The initial situation of a microsimulation more often corresponds to a representative population of the target population, so that the reasoning in terms of "what if?" allows to produce relatively "safe" measurements of the consequences of a change of a political nature. For its part, a MAS model more often allows to identify the functioning of the process behind a emergence, which often makes it start from an initial neutral situation, an empty space or a space where the elementary objects are randomly distributed.

Further on, we present three families of computer simulation models:

– random walk models in order to simulate diffusion phenomena;

– CA models with applications to the evolution of land use;

– agent-based models, while distinguishing the applications where the agents represent simple entities (individuals) or composite entities (settlements or cities).

4.3.1. *Modeling the diffusion of a migratory wave front*

The process of diffusion is intrinsically spatio-temporal: a phenomenon, such as the adoption of an innovation or an epidemic, for example, appears at a given time, in a given place and spreads progressively to other places, starting from interactions between individuals. The first to have modeled this phenomenon in geography is Torsten Hägerstrand. In his first works in the 1950s at the University of Lund, he used Monte Carlo simulations in order to study migrations and diffusion of innovations as spatial processes. In both cases, his questioning is anchored in the empirical case of Asby (the innovation involving the improvement methods of grazing in small farms), the Swedish village of his childhood [HAG 53]. Since that time, he clearly favors an approach of phenomena at the level of the individual, by grasping the logic of his/her intentions and his/her actions. He would then formalize this concern in his development of "time-geography". Since the 1950s, in his works that put forward the importance of exchanges of information on the diffusion of innovations, Hägerstrand considers that it is essential to introduce randómness in any rule that brings individuals to take a decision involving change. In the simulations that he was carrying out at that time without computers, Hägerstrand introduced this randomness using a dice within the Monte Carlo method.

The Monte Carlo method allows simulating non-deterministic phenomena. Therefore, it is frequently applied in the microsimulation models presented above. If an individual possessing a certain profile has 80% chances not to move house, 15% chances to change residence within the same municipality and 5% chances to migrate out of the municipality, a number between 1 and 100 is pulled out in a random distribution. If the number is in the [1, 80] interval, the individual is simulated as not changing residence, if it is in the [81, 95] interval the individual is considered to be changing residence within its municipality, and finally if the pulled out number is in the [96,100] interval, he/she is considered as a migrant. The operation of drawing lots is repeated for each individual constituting the sample and this for each event that has to be simulated. Moeckel *et al.* [MOE 03] use a process that falls within the same principle to proceed to the spatial distribution of a synthetic population within a metropolis. An interval of values is assigned to each cell as a function of the density of the built area. A draw is then carried out for each individual, and this is located in the cell corresponding to the number drawn. The method can also be used to simulate the choice of localization. This method is conventional whenever it is wished to introduce a random mechanism in a behavior. It has notably been used to simulate migrations.

Such a procedure was used by Young [YOU 02] to explore the colonization of empty spaces by migration. He considered a homogeneous geographical space upon which agents move around. These represent either individuals or groups (hunters–

gatherers). The author uses the Monte Carlo method to manage the two processes involved at the agent level of the model:

– the demographic change: the agents have a certain probability of dying (natural death or as a result of a conflict) and giving birth (by division if the agents are groups) at each iteration. Death may also be caused by way of congestion (if too many agents are located on the same cell);

– the migration: agents move about in space by following a random walk (eight possible directions).

The model has been used to explore the patterns of colonization of an empty space (Figure 4.7):

– if the growth rate is strong and migration rate is low, there is a front wave that moves slowly in a progressive manner. The author refers to the diffusion of the farmers during the Neolithic to illustrate this case, referring to the case of Indo-Europeans and Bantus, for example, and also that of the North-American colonization by the Europeans;

– if the growth rate is low and the migration rate is strong, there is no front. The whole space is fairly rapidly invested, but in a diffused manner and with an overall low density. The colonization of Australia during the Pleistocene is one example of such a case [YOU 02].

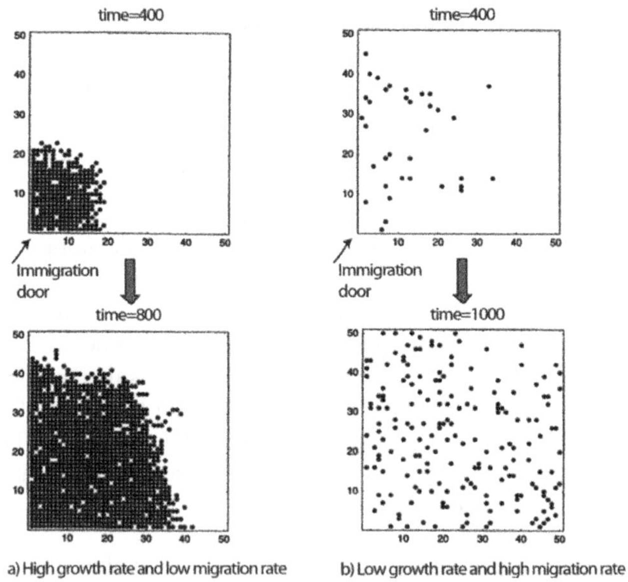

Figure 4.7. *Simulation of the colonization front according to two sets of parameters (source: [YOU 02])*

Simulating the expansion of agriculture in Europe during the Neolithic, Parisi *et al.* [PAR 08] developed a more sophisticated model incorporating a process of demographic growth and a diffusion process in which the decision to migrate depends on the state of neighboring cells. The cells are characterized by a set of parameters describing their agricultural potential, which determines their carrying capacity. The initial situation is assumed to be 9,000 BP and only one cell (assumed to correspond to South-Western Anatolia) is occupied by a population of farmers. When the number of farmers exceeds a certain threshold, part of the population migrates to a neighboring cell, chosen at random but on the condition that it is empty and that it has an agricultural potential. The rate of demographic growth in a cell is proportional to the number of neighboring cells possessing an agricultural potential. This model is able to reproduce the successive phases of the spreading of agriculture in Europe, broadly speaking. Unlike the previous model in which the space was homogeneous, the environmental constraints of Europe in this case are introduced through the agricultural potential. The authors have then traced the genealogy of the population groups and have been able to put it in parallel with the different language groups in Europe. In this example, the geography of Europe intervenes at the input of the model (through the parameters of agricultural potential) while the linguistic areas emerge, identified from the genealogy of the groups, the oldest separations being interpreted as corresponding to the strongest linguistic differentiations. The interpretation is done from the time elapsed since the different "clusters" of separation.

To make a cell change state according to the states of the neighboring cells is the basic rule of CA. Parisi *et al's* model results in fact of the coupling of a logistic growth model (see Chapter 2, section 2.1) and a cellular automaton where the cells are characterized, on the one hand, by their agricultural potential and, on the other hand, by their human occupation. This type of model has experienced a large development during the past 20 years in geography, notably in the modeling of the transformations of urban space.

4.3.2. *Cellular automata approaches: the case of land use changes*

CA is actually considered by many geographers as a geographical tool *by essence*, due to the analogy between the geographical space and the automaton representation in two dimensions with a grid. Hence, Tobler [TOB 79] mentioned a "cellular geography" and Couclelis [COU 85] mentioned a "cellular world" which allows us to reflect on and model the links between the micro- and macrogeographical dynamics. Thus, space is conceptualized in the form of a grid and each grid cell is represented by a state (living or dead in the "game of life", agricultural potential in the example mentioned above, and type of land used in many applications in geography). The change of state at time t to another at time t + 1 is governed by a transition rule relating this change to the state of the neighboring cells in t.

The first applications of CA in geography are very close to the "game of life", with two possible states for the cells (built-up and non-built area, for example), and transitional rules relating to Moore's neighborhood (eight neighbors), in order to, for example, simulate the spreading of urban growth in space. The counterpart of this simplicity is the lack of realism that it implies. Therefore, models richer on the thematic plan have quickly been developed with modifications of the original model in several directions:

– taking into account a more significant number of states. In order to simulate, for example, land use dynamics, a variety of categories have been considered in the applications (residential, industrial, commercial, green spaces, etc);

– taking into account the effect of land use, not only of the immediate neighborhood but also of more distant neighborhoods, to formalize the cell transition from one state to another. A weighting system, depending on the distance and the land use concerned, is often associated with such an approach (the effect of a commercial activity, for example, impacts a smaller area than that of an industrial activity);

– taking into account factors other than the effect of the neighborhood: the accessibility relative to the existing network of transportation, the physical suitability and the existence of planning rules.

One of the first models developed in geography in this sense is that of White and Engelen's [WHI 93, WHI 94], initially applied to simulate the development of the city of Cincinnati from its first urban core. They are interested in the spatial configuration of the three basic categories of land uses (commercial, industrial and residential), and consider in the transition rules of a land use to another several levels of contiguity to assess the effect of the neighborhood and a factor of physical suitability. The transition rule is represented there in the shape of a function linking these different factors and to which a random term is added. According to its state at time t, its location, physical suitability and the states of the neighboring cells up to a degree 7, this function calculates the potential for transformation in each of the possible states for each cell. The highest potential determines the state of the cell at time $t + 1$. Such models are rather more operational for the geographer but the consequence is nonetheless that the model's formal properties are less controlled.

This modeling framework has given rise to numerous applications and the development of a modeling platform called Metronamica [VAN 05] has led to a multiplication of works, often with an operational aim. As a matter of fact, this user-friendly tool enables the user to test the impact of a new planning, transportation infrastructure or a new urbanized area, and perform prospective. Thus, it is a tool that helps the user planify (see Engelen and Uljee for the case of the Netherlands [ENG 97], Lajoie *et al.*, [LAJ 07] for a prospective approach applied to Réunion, for example).

Stanilov and Batty [STA 10] proposed a theoretical framing for this family of models of land use dynamics by CA. They distinguished two scales involved in change:

– the local scale at which the attraction and repulsion forces operate between the different land-uses. Indeed, land uses such as residential occupations, recreational areas and commercial activities tend to attract each other. Conversely, industrial activities are repulsive for residential land use;

– the regional scale corresponding to the global physical framework, particularly in terms of transport infrastructure and organization of the network of activity centers. It determines the accessibilities to the key elements of the urban space.

The authors propose an analogy between the genetic code of individuals and the generic rules that govern the transformations of land use. This "genetic code" of the city would determine the potential urban development in the long-term. Therefore, these rules operate at all times, regardless of the economic climate. The latter has an influence, but does not come first in the explanation of change. Mechanisms such as the distance to the Central Business District (CBD) and train stations have enduring influence and play the role of the "genetic code". The economic conditions affect only afterward, in a secondary manner, creating the context in which the generic rules operate.

Stanilov and Batty apply their CA model, using the platform Metronomica, to the evolution of land use in West London over a period of 130 years. The years 1875, 1895 and 1915 are used for the calibration and then forecasts are made and compared with the states observed for the years 1935, 1960, 1985 and 2005. The authors stress upon the overall good quality of model predictions for these dates. They conclude that a small number of variables are thus sufficient to give account, in general terms, of the spatial organization of land use transformations on the long term. The model has notably reproduced the transitions from a nucleated type of growth (with separate urban cores), to a diffuse type, and finally to an infill type over a century. The ability of the model to reproduce a realistic change over such a long period corroborated, according to the authors, the hypothesis that the rules defined at the local scale, which govern the changes of land use, are consistent through time and independent of the socio-economic context.

In a general fashion, CA have given rise to many research directions in the field of geography. Some are theoretical, relying, for example, on couplings with microeconomics models. Others are operational and often coupled with GIS [POR 97, BAT 05, BEN 04]. We propose to quickly introduce a few of these experiences in order to illustrate the diversity and richness of the research possible in this area.

Caruso *et al.* [CAR 07] integrate a theoretical model of urban economy with a CA to explore the conditions for the emergence of a mixed belt between the center of the city (CBD) where employment is concentrated and the countryside occupied by agriculture. This mixed belt at the periphery of the center is composed of residential areas, more or less dispersed, and agricultural areas. The spatial structure which emerges, rather denser in the central area than on the outer fringe, shows some stability in time. The model is of theoretical nature and corresponds to the case of the monocentric city, with an isotropic space at the center of which is located the CBD.

The mechanisms introduced in the model rely upon the results of many econometric models applied to a diversity of cities and based on statistical models (multiple regression and logit model notably) where the variable "to explain" is, for example, the value of real estate property (hedonistic model) depending on the characteristics of that property (for example, for a house: the number of rooms and bathrooms) and the environment in which it is located (proximity to green spaces, a train station, for example). These works show notably the attractive effect of green spaces, whether they are parks or peri-urban agriculture. This aspect is integrated in the model through the inhabitants' preferences. Each cell is occupied by either a residence or agriculture. The economic model evaluates the choice of new migrants on the basis of arbitration between the cost of transportation to the CBD and the attraction for the neighborhood. Two components are involved to characterize this neighborhood: the social amenity and the presence of a green environment. It is a competitive market where the cell is assigned to the highest bidder. The installation of each new migrant transforms the neighborhood and, thus, the conditions for the following migrants. The simulation results show that the degree of diversity of the urban space depends on the relative importance accorded by the agents to the social amenities relative to the green amenities. The mixed belt emerges at the periphery of the compact urban center when the preferences for green spaces are accentuated [CAR 07].

Another research direction looks to develop a more empirical approach, in particular, by seeking to adapt the CA to a realistic geographical space. Putting forward the nonlinear perception of distances, van Vliet *et al.* [VAN 09] extends Engelen and White's work by proposing scalable neighborhoods, the size of the spatial units qualified as *neighboring* varying as a function of their distance to the cell for which transition is being considered. In other works, it is the geometry of the cells itself, with the construction of polygons with varied shapes, that is adapted to the empirical cases being studied. Such is the case of the models developed by Benenson *et al.* [BEN 02] and Moreno *et al.* [MOR 09] to simulate the emergence and the evolution of a segregation process, in the case of Yaffo for the first, and of Pau for the second. Used in the environmental field, CA have also been adapted in order to be applied to empirical cases. Thus, Douvinet *et al.* [DOU 08] use the RuiCells model to study the influence of the morphology on the hydrological behavior of watersheds. In this model, the transition rule simulates the displacement of a stream from one surface to another, depending on the morphology of the terrain that creates a "roadway" following the principles governing liquid flows (water or mud). This model has been used to construct, from simulations, synthetic indices of the potential dynamics of a watershed [DOU 08].

Finally, the CA are often coupled with other methods. Already in the 1990s, Couclelis [COU 97] stressed upon the benefits of coupling CA with geographical information systems (GIS). Some environments of development of CA are coupled with GIS and incorporate their leading features, which makes it possible to import the geography of observed situations for designing the initial state of the studied system. This coupling also helps to evaluate the different states produced during the simulation by comparison with the states of the analyzed territory. In this case, the applications have operational objectives. This is the example developed by Dubos-Paillard *et al.* with the SpaCell automaton (©Langlois[9]) [DUB 03], to analyze the urban evolution around Rouen between 1950 and 2000. Three maps are integrated in the CA: 1950, 1966 and 1994, that describe the land occupation in 15 categories on a grid of resolution of 150 m. The authors identify two distinct processes of urban growth: densification over the first period, and urban sprawling over the second period. The CA is developed with the objective to reproduce as precisely as possible the observed spatial organizations and, thus, bring out specific types whose behaviors escape the logic of change integrated in the CA. This is the case of some major public housings, for which the CA does not reproduce the implementation.

Phipps and Langlois proposed, on their part, coupling the CA with an aggregated dynamic model [PHI 97]. Engelen *et al.* [ENG 99] combine these two forms of coupling by inserting their initial CA model between a GIS providing the initial situation of the land use at a fine scale and a dynamic model at an aggregate level managing the change of land uses at the macrogeographical level. Once the total number of cells of a certain land use is known, the geographical distribution of this stock is then governed by the CA.

9 Langlois P. http://www.spatial-modelling.info/cellular-automata-spacelle.

Thus, CA offers a particularly powerful framework for modeling spatial dynamics when change depends on what is happening in the neighborhood, following the famous Tobler's law[10]. However, when the driving interactions of change obey other rules with, for example, an effect of hierarchy, an agent approach may appear more appropriate. Therefore, the coupling between CA and MAS represents a path of fruitful research. Hereinafter, we will distinguish the MAS applied to *simple* elementary entities (for example, individuals), to recall a categorization presented in Chapter 1, and the MAS based on agents representative of *composites* elementary entities (case of cities). In both cases, the concern is about modeling elementary entities, in the sense that they are indivisible relative to the problems focused on (individuals of a population, in the first case, and cities of a system of cities, in the second case).

4.3.3. *Agent-based models applied to simple entities*

A number of agent-based models developed in geography and archeology are designed to explore what properties and structures are emerging at a meso- or macrogeographical level from the interactions operating between individuals or households. These interactions may be direct (information is exchanged between two agents) or indirect, if they operate through the environment (an individual who changed localization transforms, with this action, the environment of other agents). The three examples which will be presented illustrate the diversity of ways in which the interactions can be taken into account in agent-based models. The first example is the more spatialist one. It formalizes the effects of neighborhoods on the evolution of the spatial configuration of different populations within the city. The second example proposes to consider the effects of interactions of different natures, some between the agents of the same social group, others between the agents and the aggregate level where the composition effects are observed.[11] In the third case, the driving interactions focus on the relationships between the agents and their physical environment. The interactions between agents in this case are indirect. In all three cases, empirical data have been mobilized and the challenge is to identify the rules operating at the elementary level of individuals or households, which lead to the emergence of changes similar to those that have been observed. These applications are related to changes whose time scales range from a few decades to several centuries.

10 "Everything is related to everything, but near things are more related than distant things" [TOB 76].

11 Composition effect is the effect of the aggregated characteristics of a population. Thus a characteristic can have an effect at the individual level (competence, for example) and have another effect at the aggregate level by composition of the different values of this characteristic.

4.3.3.1. *Population segregation dynamics in the city of Yaffo between 1955 and 1995*

This example was mentioned earlier to discuss the form that the outputs of MAS models can take. Starting from the distribution of Jewish, Christian Arab and Muslim Arab populations in 1955, the objective of the authors is to show that it suffices to formalize the preferences of individuals with respect to residences from two points of view, in order to reproduce the redistribution of these populations over the next 40 years. These two points of view are:

– the composition of the neighborhood on the cultural plan, making attractive, for a given individual, environments in which individuals of the same cultural profile as him are majority;

– the buildings' architectural style according to two types: the oriental style preferred by Arab Israelis and the style by "blocks" preferred by the Jewish Israelis.

The model has been tested with different parameter values giving differentiated weights to these two factors. As previously discussed, the outputs of the model are apprehended at the level of the city of Yaffo, through two indicators: the share of the Arab population in the total population, simulated at different dates, and Moran's spatial autocorrelation index, indicating the segregation degree of the population (that is the degree of spatial concentration of the population of same cultural type). If the parameters associated with these two factors characterizing the preferences are null, the share of the Arab population at the final date is largely underestimated. The same happens when only the parameter associated with the cultural neighborhoods' effect is not null. Only when the parameter associated with the architectural style is not null, is Moran's index not very high, resulting, at some periods, in cohabitation between Jews and Arab Israelis in the same buildings, a result which contradicts the empirical observations. The authors conclude that the only simulations correctly reproducing, at the same time, the developments of the share of the Arab population in the total population and Moran's index, are those for which the parameters associated with the two indicators are not null. The two parameters have to be equivalent and relatively low in order to get simulated indicators close to the observed indicators [BEN 02]. The authors emphasize particularly that it has not been necessary to take into account the price of real estate as a constraint in residential choices to obtain evolutions close to what has been observed.

4.3.3.2. *The "spiral" of school inequalities*

Interested in the genesis of the social differences in terms of diplomas, Manzo [MAN 09] developed a MAS model formalizing the individual decisions of students and their families about the education level that they have as a goal according to three families of hypotheses:

– a rational dimension, leading to choose according to the respective profitability of different educational levels considered, these being perceived differently depending on the social environment of the individual;

– an interaction dimension that will make the choice of the individual dependent on its exchanges with other agents within the scope of its contact network, by imitation;

– an effect of the interaction between the individual and the aggregate level of the school. Indeed, the social profile of the school is determined by its composition which has emerged from the individual choices. In turn this profile influences the choice of the students.

Thus, the first two dimensions are related to the individual level, on the one hand, its own rationality and, on the other hand, the imitation of persons from his network. The third dimension refers to the fact that the social aggregates (here the social profile of the school's pupils) have a retroactive effect on the action of the agents.

The simulated data are then analyzed by means of statistical processing.[12] This step allows to put forward the type of educational stratification that emerges from the rules formalized by the MAS. The effect, notably, of the importance of the interactions between agents of the same social group is explored. The confrontation with the empirical data observed in France and Italy shows that in these two cases the configuration of these interactions played a key role in the educational stratification that is observed there.

4.3.3.3. *Simulating the abandonment of a territory: examples in archeology*

In the field of archeology, different research teams have used MAS in order to study the population dynamics of environmental change in the long run. The objective here also consists of creating an artificial world in which households evolve when confronted with a certain environment, where objective is to extract the resources necessary to live.

An emblematic case concerns the Anazasi in the "Long House Valley" in Northeast Arizona between the years 800 and 1350, the date at which the valley was abandoned [AXT 02, GUM 03]. The challenge is to simulate the population evolution during this period, as it has been reconstituted by the archaeologists from different vestiges. This evolution shows two peaks, one peak toward 1100, the other a little before 1300, and then a collapse toward 1300, followed by an abandonment of the valley. Environmental data have been reconstructed for the whole of this

12 It should be emphasized here, that the same statistical tools actually make sense, whether the data is derived from the empirical observation or whether it results from simulations.

period, which allows the authors to estimate the agricultural potential at each place of the valley. Each household has a demographic functioning, nutritional needs and a capacity for agricultural work which are implemented in the model. The central rule is the one that manages the migration of a household. This occurs if the stored crop added to the year crop is inferior to the nutritional needs of the household. The household then chooses a new plot; one that can fulfill its needs. If such a plot does not exist, the household leaves the valley. The first simulations do not manage to reproduce the population evolution curve. This evolution will only be correctly reproduced (Figure 4.8) with the condition to introduce heterogeneity in the demographic and nutritious behaviors of households.[13] However, no simulations result in the abandonment of the valley: the environmental conditions are actually such that a small population could be maintained. Therefore, the explanation is to search in factors that are not taken into account in the model, of social and/or anthropological order: below a certain population size, the community has taken the decision to leave the valley [AXT 02, GUM 03].

While recognizing the validity of this model, some authors point out that the same results could have been achieved by developing a model more parsimonious at the aggregate level [JAN 09]. However, the interest of this example is to illustrate the driving role that the interindividual diversity can play, an epistemological position often defended by the microsimulation modelers.

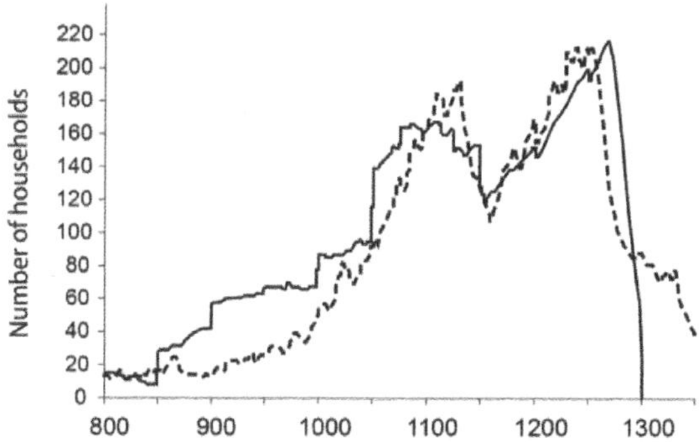

Figure 4.8. *Population curves of Anazasi households [AXT 02]*

13 An example, to clarify: in the first version of the model, a child in a household separates from his family and forms a new household at the age of 16. In the second version, this action corresponds to a random selection from a normal distribution of mean 16, i.e. there will be heterogeneity in the ages of moving out. In the same way, the size of households and their nutritional needs are supposed to vary.

Another example relates to the modeling of the Pueblo population in Mesa Verde, in Southwest Colorado, between 600 and 1300 [KOH 07, KOH 08]. In this case, fine-grained information has also been reconstructed detailing the environmental resources in terms of agricultural potentials (corn), water and hunting (cervids); these resources vary ever year according to climate variations. The agents make rational decisions, such as minimizing their caloric expenditure by maximizing their proximity to the resources. It appears that the first simulations underestimated rather systematically the evolution of the population: it appeared to be limited, in the simulations, by the strong decrease of cervids, a consequence of hunting to cover the protein needs of the population. This deviation between archaeological estimates and simulation outputs led the authors to formulate a new hypothesis that involved the conception of turkey domestication. This domestication probably replaced the protein source that the cervids represented.

The authors have developed a family of models, gradually introducing more complete interactions, for example exchanges between households when one of them has suffered a setback when harvesting, following a principle of reciprocity. The simulations have been able to reproduce a concentration process of the settlement with formation of villages greater in size. Alternating between the simulations' results and archaeological reflections has enabled to show that the abandonment of the region was the result of a combination of climate deterioration and conflicts. In this example, the MAS is a tool to help reflect on a thematic issue, the understanding of the functioning of the Pueblo societies and of the factors that cause them to abandon a territory. In this example, as in the previous one, the MAS has enabled to show that the classical explanation based on a simple causality of climate change was insufficient.

4.3.4. *Agent modeling on composite entities*

As we emphasized earlier, agents can represent entities other than individuals (persons, animals, etc.), that is to say aggregated or composite entities, particularly groups (hunter-gatherer, for example) or settlements (hamlets or cities). We illustrate such a choice with two examples: one concerns the fall of the Maya civilization; the other, the dynamics of the European system of cities, to which we will dedicate more attention. In both cases, the assumption is that the settlements are indivisible entities and the driving interactions of the systems involved are defined at the level of these settlements, as nodes in a network of exchanges, and not at the individual or household level as in the previous examples.

4.3.4.1. *The MayaSim model: simulating the dynamics of the Maya settlement*

MayaSim [HEC 13] models a socio-ecological system linking different environmental dimensions (climate, soil and hydrology) following an approach similar to that developed by Kohler *et al.* mentioned above, as well as demographic, social and economic dimensions. The objective is to identify the conditions of resilient or non-resilient developments of such a system, characterized by the interconnection between variables with different dynamics, some slow (land degradation), others fast (commercial connections and land use) and others, intermediaries (demography and climate). The author writes explicitly that he is looking for "candidate" characteristics to explain whether a system is or is not resilient. The model is used as a laboratory experiment to better understand, on the one hand, which dynamics have led to a dense and well interconnected population of the Maya and, on the other hand, how this socio-ecological system has responded to the change in environmental conditions.

The model is based on a combination of several methods: MAS, CA, GIS and a network model. The space is divided into 20 km^2 cells. GIS is used to manage the empirical data about the environment, which are used to estimate the agricultural potential and the access to water resources at the cell level. The agents represent the settlements and MAS handles their interactions with their environment (agricultural activity and food consumption) as well as the consequences of these interactions in terms of demography and migration. AC is used to formalize the development of the forest. At last, the agents are connected in networks within which they trade. From this trade they make profits. When a settlement exceeds a certain population threshold, a mechanism allows to expand the road network to settlements located within a radius of 40 km, facilitating trade with them.

The simulations are spread over 600 time steps, starting from an initial situation without settlement, reaching a maximum expansion toward 400 iterations, followed by a decline. The system shows great sensitivity to the interaction between the growth rate of trade, which generates increasing profits, development of the population and land degradation. Around the 400th iteration, the system is at its peak: all the settlements are interconnected, the population is large, the agricultural area occupies a large part of space (at the expense of the forest which has receded), and the land is degraded. At this stage the system is very vulnerable and climate variations will result in strong demographic effects. The decline of a major node in the network will result in a cascade of declines. However, all the simulations do not lead to a strong increase of the population and do not lead to a peak corresponding to the estimate made from the observation of the archaeological remains. The more determining parameters are the ones managing the trade growth and land degradation.

The author concluded that his model helped to demonstrate that there is not just a single causality to explain the decline of the Maya civilization (a drought has sometimes been referred to as a possible explanation). It is the interactions between several phenomena (climatic variation, population pressure, deforestation, soil degradation and configuration of commercial networks) that determine if the system is resilient or not. The author notes that such interaction effects are specific to a complex system and that the MAS formalization is particularly adapted to explore it [HEC 13].

The comparison between this model and those of the Pueblo and Anazasi populations, mentioned above, is interesting. It illustrates that a same problem, that of understanding/explaining the abandonment of a geographical area, can be formalized by making different choices about the elementary entity, the households in the case of the works of Kohler *et al.* on Colorado's Pueblo as well as those of Axtell *et al.* on the Anazasi of Arizona, and the group level for the MAS developed by Heckbert on the Maya. The first two favor a positioning in terms of methodological individualism, and formalize the rational behavior of households faced with their environment. The interactions that operate at this level are considered as the driving interactions. In the case of the Maya, it is the interactions between the groups, at the level of their settlement, which are regarded as being the drivers.

In Kohler's as well as Heckbert's works, the concept of "network" is mobilized. The set of exchanges within networks is assumed to be essential in the dynamics of the settlement systems in both approaches. However, the *level* considered to formalize the action is different. The following example allows us to explain further how to formalize networking within a settlement system, this time in the case of a system of cities.

4.3.4.2. *The EuroSim model: simulating the dynamics of the system of European cities between 1950 and 2050*

The objective of the EuroSim model [MAT 08, PUM 09, SAN 10] is to explore the role of trade among cities in their respective evolutions over a period of a century, from 1950 to 2050. The initial situation corresponds to urban Europe in 1950, which is already highly structured. The objective is to identify the mechanisms necessary to ensure that the urban structure in place is maintained or, on the contrary, transformed (transition to a more polycentric organization, for example).

The "city-agents" have attributes (population and economic specialization) and have two roles. On the one hand, they are producers of goods and services that will be offered within their networks of exchanges, on the other hand, they are consumers of goods and services produced by other city-agents. At each time step, each city-agent constitutes a network of exchange for each of its economic specializations. The construction of these networks in three stages is one of the distinctive features of the approach:

– First step – *the potential exchange network* is composed of cities with which an exchange could take place given the specialization of the city. These are cities in close proximity in the case of the regional functions, or cities with strong functional reciprocity in the case of specialized functions, such as finance.

– Second step – *the information exchange network* includes a selection of cities among the previously mentioned. Following a double principle of exploitation/exploration [MAR 91], consisting of valuing what is already acquired and exploring in parallel new ways to innovate, the network is made up of part of the city-agents that emerged as the best customers during the previous steps and a set of cities randomly drawn inside the potential network. The cities being, simultaneously, producers and consumers of goods and services, a multitude of entwined exchanges take place.

– Third step – *the network of effective transactions* emerges from the set of information exchanges between the city-agents corresponding to the second step. Each city agent looks to sell its production to satisfy its demand in a context where exchanges are intertwined, reflecting the competition between the cities. The effective transactions are reflected by a transfer of wealth.

The period 1950–2000 was used to calibrate the main parameters of the model by comparing the values of simulated and observed macrogeographical indicators (evolution of the total European population and its rank-size organization in particular). An informal sensitivity analysis[14] has been made through numerous simulations. Scenarios were then developed to explore the effects of two kinds of policies on the evolution of the European urban system up until 2050: (1) according to the degree of opening of the European borders with regard to the immigration from outside Europe; (2) according to the existence or non-existence of economic barriers between the blocks of the Eastern and of the Western European countries [MAT 08]. Therefore, four "extreme" scenarios have been defined as:

1) a double opening, defining a context of demographic growth and free economic exchange between Eastern and Western European cities;

2) an opening to immigration, with the associated demographic growth but within a context of economic closure between Western and Eastern Europe;

3) a closure to immigration, with the associated demographic decline, but a free economic exchange between Eastern and Western cities of Europe;

4) a double demographic and economic closure.

Figure 4.9 illustrates the ability of the model to produce different trajectories: the artificial city (*L*), a great metropolis with many specializations at the international

14 A large number of simulations corresponding to different sets of parameters have been carried out, but these simulations have not given rise to systematic replications.

level, and the city (*G*), smaller and only with industrial specialization. These respond differently to the existence of economic barriers between the Eastern and Western blocks. While the trajectory of the first city is not influenced by these barriers, the second city is limited in its growth within the appropriate scenario. Having multiple specializations, and hence being integrated in several exchange networks, the first city (*L*) resists better to the "closure" scenario than the second city (*G*) which is monospecialized. Such a difference in behavior corresponds to the empirical observation, where cities with a diversified economic profile are often more resilient, although no such rule has been explicitly introduced. A follow-up of the different cities' exchange networks shows that cities respond differently to the existence of economic barriers according to their size and economic specialisation. The exchange networks of a large and multispecialized city tend to expand when there are no economic barriers (which is consistent with the intuition) while the opposite is true for the smaller and monospecialized cities that suffer greater competition in this same situation.

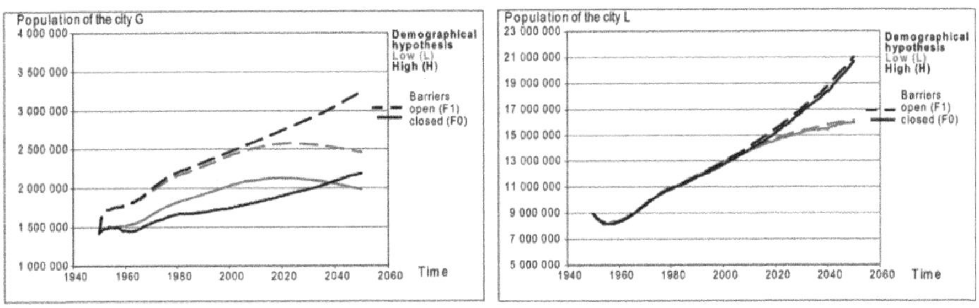

Figure 4.9. *Demographic trajectories in reaction to demographic and economic scenarios*

Finally, at the level of the system of cities' structure, the simulations show that, regardless of the set of parameters, the rank-size organization remains extremely stable over time. Indeed, whether the immigration policies are favorable to a Europe that is open or closed, or whether there are any constraints on economic exchanges between the cities, the simulated hierarchical organization of the system of European cities remains the same in the long-term.

The changes that emerge at the level of the global structure of the system are not contained in the rules governing the interactions between the city-agents. This stability in the rank-size organization reflects, as it has already been pointed out in Chapter 3, the existence of different logics of exchanges that are inserted in a well-integrated global system. The EuroSim model allows us to confirm the stability of this structure at the level of the system of cities, while making it possible to identify some recompositions at the level of the cities and their interactions. At that level the responses to change are much more diversified.

The analogy mentioned by Stanilov and Batty [STA 10], between the fundamental rules governing the evolution of a city's land use and a "genetic code", would also be fully meaningful for the EuroSim model. During a period as long as a century, the economic and technological contexts are profoundly modified (new economic functions emerge, the travelling times change, etc.), but the driving force of the system of cities' dynamics is the same: each city seeks to enhance its relative position in the urban system through the interactions with the other cities belonging to its different networks of exchanges. As a matter of fact, it is the success of these interactions (in terms of trade) which secures its place in the urban hierarchy, or even makes it move up in this hierarchy.

4.4. Conclusion

The methods presented in this fourth chapter have in common the fact that they are intrinsically dynamic and are mobilized in the context of an explanatory approach in social sciences. The aim is to formalize the process that leads to change, the focus being on: segregation (Schelling's model, Yaffo's model and school inequality model), growth, concentration and hierarchization (models of the Pueblo of Mesa Verde and Eurosim), abandonment (models of the Maya, the Anazasis and the Pueblos) and diffusion (land occupation and periurbanisation). The modeling formalisms are varied and the preference for one rather than the other depends on the scientific habits, the objectives and the hypotheses of the mechanisms involved in the processes at stake. The objects, attributes and relations are then identified in a manner consistently relative to the problematics and the chosen method:

– the CA approach will be adapted if the hypotheses can be formalized in terms of cells that change state depending on the state of neighboring cells;

– the microsimulation: if the consistency of family construction and demographic aspects play a key role in the evolution of the system;

– an MAS: if a direct or indirect interaction with other agents affects the behavior of an agent.

Indeed, interactions and driving forces of change are of a different nature depending on the type of models. The applications show a clear difference between microsimulation on the one hand and MAS and CA on the other hand. In the applications of microsimulation in social sciences, the interactions mainly refer to the dependence of the decision of some, to that of the others:

– the migration of one household member has a high probability to cause the same for the other members;

– the migration of a household makes a house available on the real estate market, etc.

In the CA models, the interactions are made in the form of influence, following a logic of spatial diffusion of a phenomenon. The MAS models present a very wide variety of applications:

– in the case of mobile entities, the interaction is reflected by a meeting in the space, which can result in the transmission of information and can be used to formalize a diffusion mechanism. Otherwise, it can be reflected by a competition to occupy a given space and/or exploit resources;

– in the case of fixed entities, the interactions are expressed in the form of message exchanges;

– the interactions can also be indirect, where the choice of an agent to occupy a place (a housing or an agricultural parcel) creates a change in context for another agent (in the same way as seen in the examples mentioned above with regard to microsimulation).

Another difference between these families of models lies in their objective and also in the relation to empirical reality. In the case of microsimulation, the operational dimension of the model is most often favored and the aim is then to test the effects of different scenarios. The approach is of the "what if?" type: how will the population be redistributed if the price of oil increases? In the case of climate change of such magnitude? In the case of the arrival of a significant flow of refugees? If a new shopping center is created at such a place? In these models, the transitions of individuals from one state to another are most often governed according to statistical principles, based on the covariation of demographic and socio-economic variables. Most MAS are instead developed in a perspective of supporting reflection and the interest is very specific to the *mechanisms* at the origin of change. The purpose is to bring forward the emergence of structures or changes to structures and understand the underlying logics. The practice is more often to compare the consequences of different hypotheses of the processes in play on a system's evolution than to study the effect of exogenous events, even if this is entirely possible from a theoretical point of view[15], once the model is stabilized and valid. However, microsimulation and MAS offer different scientific contexts to explore the effect of exogenous events; in both cases, the model is manipulating the evolution of an artificial population. In the first case, this population is described in

15 Let us point out that in a number of cases, the differences are more likely a matter of practices and habits than of constraints imposed by the formalism.

a complete manner from the point of view of attributes and statistical relationships, whereas in the second case, the agents are in general much more stylized. However, in this latter case, the mechanisms of change are explicitly approached following a generative principle [EPS 07]. That is to say, as it has been pointed out above, the model makes explicit the rules which allow "generating" the structure to which the system results. In order to characterize the structure of the system at the final step, the concept of *resultant* would, thus, be more appropriate in the first case and that of *emergence* in the second case.

Finally, the relationship with the empirical reality also differs from one application to the other using MAS. Therefore, the literature confronts the models known as *Keep It Simple, Stupid (KISS)*, simple and stylized (Schelling's model is a typical example), to the models known as *Keep It Descriptive, Stupid (KIDS)* favoring a more inductive approach mobilizing all the factors involved in the evolution of a phenomenon (the example of the Pueblo in the Mesa Verde presented earlier would be an example) [AXE 97, EDM 05, BAN 13]. Here again, we prefer to emphasize the reciprocity of the two perspectives, one that allows us to bring forward the generic mechanisms in the change, and the other to give their place to the complex interactions operating in the empirical world.

Conclusion

Each of the points mentioned in the book has generated a vast literature. The fields involved are changing rapidly and are still subject of scientific discussions. The approaches and methods exposed result from multiple interdisciplinary dialogues and from the gradual integration of concepts drawn from different fields. Among the cited interdisciplinary dialogues, some have not always been in favor of spatial analysis. For example, before the emergence of a trend of statistical methods incorporating dependence and spatial variability, geographical data were first considered unsuitable for statistical processing. Furthermore, during the early days of the development of geomatics, dialogue with spatial analysis was not as evident as it is today. Geographical information systems (GIS) have facilitated the management of geographical information in processing, but often with a very operational objective. This has not motivated the development, in GIS, of models to follow up spatial data in time [OSU 05], or of functionalities that allow developing models in spatial analysis [FOT 00]. Today, the tendency is, on the contrary toward an integrative co-evolution. Geographical information science is integrating a variety of methods of spatial analysis and is often coupled to dynamic modeling.

The choice has been to focus on the interdisciplinary interfaces. Rather than going deeper with specific methods, we have used this position to enlighten epistemological and methodological questions throughout the whole chain of modeling. The bibliographic references are therefore numerous in order that each and every one can improve the more specialized or specific methods. In fact, the whole coherence of the approaches has been favored over the more operational aspects. This is the case for the modeling of spatio-temporal data, as well as for the statistical methods, the spatial analysis models and the simulation models.

We have greatly insisted on entity construction and meaning, especially in the first two chapters. The reflection on these points is essential when facing composite entities, whose definition can be varyingly complex. Such an approach provides an alternative to a recurring position of methodological individualism when using multi-agent systems (MAS) in social sciences. Indeed, most of these applications mobilize in a privileged way the simplest elementary entities, notably individual beings or households, whether in human geography or in archeology. For Benenson and Torrens, such a choice allows escaping the problem of the modifiable areal unit problem (MAUP), which makes the results (of statistical model or gravity model, for example) dependent on the definition of the entities on which the treatment operates [BEN 04]. Formalizing the rules of evolution at the agents' level corresponds to an epistemological positioning of methodological individualism. The interactions between simple entities (at the most basic level; here the people) may however not correspond to the same principles as the interactions between the composite entities that they are part of (whether they are social or spatial aggregated entities). From a methodological point of view, MAS corresponds to an individual-centered approach but this does not mean that the epistemological framework is that of the methodological individualism. An individual-centered model formalizes the behavior of elementary entities rather than that of the aggregation of these elementary entities, whether these elementary entities are: individuals, houses, firms, but also parcels, settlements or cities. Methodological individualism, for its part, systematically refers to individual beings in the sense of economic or social agents as the elementary entities of an analysis. The MayaSim and EuroSim models presented in Chapter 4 are hence methodologically individual-centered models, but the elementary entities are composite (city-agents and group-agents) and the approach does not relate then to methodological individualism.

A number of discussions on this issue are conducted in economics and sociology. Lars Udehn [UDE 02] therefore proposed to explicitly distinguish the approaches of methodological individualism depending on whether they are targeting the definition of social concepts, the explanation of social phenomena or the reduction of social laws into individual laws (physical or psychological). These epistemological nuances would allow going further in the construction of the explanation of spatial phenomena, or at least giving a complementary viewpoint to that given by most microsimulation models or MAS models developed at the individual level. The elements of Chapter 1 allow us to conduct a reflection in this direction to formalize in a pertinent manner the links between different geographical levels: the examples discussed in Chapter 4 illustrate the interest of using MAS to simulate the effects at a higher level of organization, of interactions operating at a lower level of organization. Two types of examples have been mentioned:

– the effects of interactions between agents representing human individuals on the spatial configuration of a territory (typical example of Schelling's socio-spatial segregation model);

– the effects of interactions between city-agents on the spatial and hierarchical organization of a settlement system (e.g. EuroSim).

However, it is not known how to build a model simultaneously integrating the three levels involved (those of individuals, cities and systems of cities), and the two nested stages of emergence associated. The main reason is the semantic change of the interactions when one passes from one level to the other: the interactions between territorialized entities relate, as a matter of fact, to another subject than those between individuals. How to identify spatial entities which make sense in the segregation field that has emerged at the mesogeographical level, in the case of Schelling's model, for example? Or in the settlement pattern which has emerged from the interactions between Pueblo households? And how to formalize interactions between groups without these groups being identified as autonomous entities? These difficulties are related to the definition, identification and delimitation of entities of interest that emerge from the interactions between more simple entities.

In the field of data processing, these issues of entities' definition are also fundamental. If we have not mentioned in Chapter 3 the question of the MAUP, crucial in the dialogue between statisticians and geographers, that is because our position in this regard coincides with that of Openshaw: "according to one given domain, if geographers would agree on what constitute objects of research and if these objects would be defined in a non-arbitrary way, the problem of the spatial aggregation would disappear" [OPE 81]. When the meaning of the entities is well specified, the possible differences in results obtained by applying the same method to two sets of spatial entities delimited differently, rather than being a problem, reveal the multi-scalar structure of the geographical space [MAT 06]. These issues of entities' definition are today reactivated with the new data produced by all the sensors that surround us, RFID chips, RSS feeds, social networks exchanges, geoweb2, etc. All these data (operational data, declarative data, etc.) constitute a crude material for which the elementary entities are rarely the entities to be analyzed: the use of this material needs in a first time to filter the data in order to reconstruct entities (simple or composite) having a meaning for the question that was raised. The concern is, for example, to identify stopping places from mobile phone data and use it to study individual's spatial practices [OLT 12]; to use regions cited by AFP (Agence France Press) for following a diffusion, etc. The use of an approach based on an explicit ontology takes its full significance here.

The position taken in this book has been to browse all the steps associated with a spatio-temporal problematic, from the critical identification of the concerned entities to the simulation of the processes underlying the evolution of the phenomenon of interest. The approaches described in the different chapters most often mobilize different scientific communities. The point was, and it is the originality of the proposed approach, to include these steps in a chain, or rather in a methodological system. Even if all the steps are not necessarily to be improved in any work, it is useful to be aware of those that are put *in brackets*. The constitution of a spatio-temporal database on a new phenomenon (or at least that which has not been specified), for example, is in itself an ambitious objective. Queries are, in this situation, a good tool for representing and describing the phenomenon. However, it is interesting to reflect, already at this stage, on the questions likely to arise in terms of interrelations, typology, and about the processes that have led to the changes observed for the phenomenon in question.

The relationships between these stages are of several natures, involving both concurrence and complementarity:

– given the objectives, an approach and a method may appear more appropriate than others;

– an approach may be preferred over another according to the habits and competences of the researcher;

– the different stages can be understood in their complementarities and coupled in different ways.

Following the exposition order chosen in this book, Chapter 4 proposes methods located at the bottom end of the research process: the entities have been constructed and have given rise to a critical discussion (Chapter 1). They have been designed and modeled while integrating time, the interrelations between the different phenomena involved and the form of the interactions between the places have been described and explored using queries and statistical methods (Chapters 2 and 3). These explorations have enabled us to develop a number of hypotheses about the processes that have led to empirically observed evolutions. The simulation models, finally, make formalizing these hypotheses possible and checking their relevance and consistency (Chapter 4). The prime example of Chapter 2 illustrates such a sequence, whose degree of genericity can be evaluated by confronting the phenomenon being studied in other spatial or temporal contexts (that is a similar system situated in other places or at other times). This work's description, however, is very stylized and the reality of the research rarely follows such a linear plot. If the challenge is about bringing forward the causes of a change in an area already well-known beforehand (from the literature and the empirical work carried out elsewhere or in other contexts), the researcher will be able to try, for example, to directly develop a MAS model. During the confrontation with the empirical data, some results of such a model will cause reinterpretation of the meaning of the entities involved (refer to point 1 in Figure I.1 of the introduction) and/or to mobilize statistical treatments to understand the meaning of some empirical regularities. Furthermore, the simulation generates large quantities of spatio-temporal data, whose exploration will be carried out by mobilizing the approaches outlined in Chapter 3. Going back and forth between points 3 and 4 can thus be done in multiple ways:

– to return to an empirical exploration of the phenomenon being studied;

– to explore the results of the outputs of simulation;

– to confront empirical and simulated evolutions of the phenomenon being studied, and this most often at multiple scales (at least the level of elementary entities and that of the system as a whole).

As illustrated in Figure C.1, the *interactions* between reflections, concepts and methods presented in the four chapters are multiple. Thus, rather than representing the succession of these different phases in a linear way, it seems to us that the spiral is more accurate for illustrating the climbing up on the path to kwowledge building while always keeping an overview of the methodological landscape that has been built.

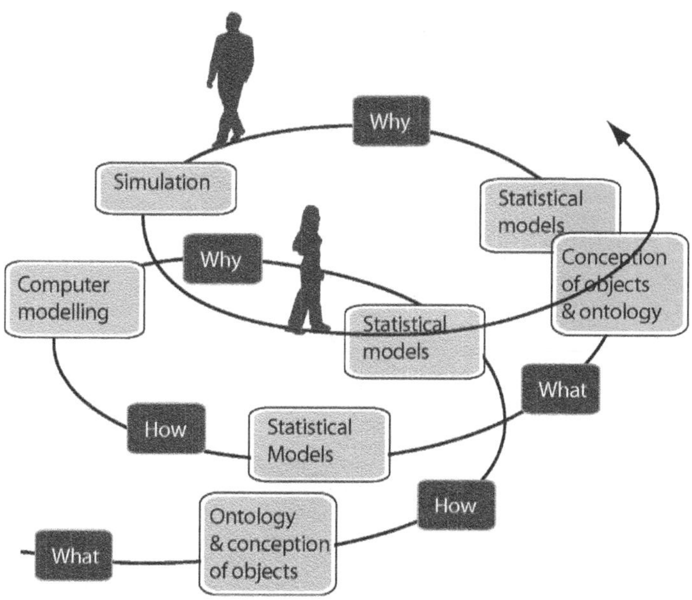

Figure C.1. *Approaching spatio-temporal phenomena: a spiral of stages*

Bibliography

[ABR 99] ABRAHAM T., RODDICK J.F., "Survey of spatio-temporal databases", *GeoInformatica*, vol. 3, no. 1, pp. 61–99, 1999.

[ALF 02] ALFREDSSON E., Green consumption energy use and carbon dioxide emission, Doctoral Thesis, Department of Social and Economic Geography, Spatial Modeling Centre, Umeå University, 2002.

[ALL 97] ALLEN P., *Cities and Regions as Self-organizing Systems; Models of Complexity*, Gordon and Breach Science Publishers, Amsterdam, 1997.

[AND 03] ANDRIENKO N., ANDRIENKO G., GATALSKY P., "Exploratory spatio-temporal visualization: an analytical review", *Journal of Visual Languages & Computing*, vol. 14, pp. 503–541, 2003.

[AND 05] ANDRIENKO N., ANDRIENKO G., *Exploratory Analysis of Spatial and Temporal Data: A Systematic Approach*, Springer-Verlag, New York, 2005.

[ANS 95] ANSELIN L., "Local indicators of spatial association – LISA", *Geographical Analysis*, vol. 27, pp. 93–115, 1995.

[ANT 06] ANTONI J.-P., "Calibrer un modèle d'évolution de l'occupation du sol urbain. L'exemple de Belfort", *Cybergeo: European Journal of Geography*, Systems, Modeling, Geostatistics, document 347, 2006. Available at http://cybergeo.revues.org/2436.

[ARC 98] ARCHAEOMÉDES (DURAND-DASTES F., FAVORY F., FICHES J.-L., *et al.*), *Des Oppida Aux Métropoles*, Anthropos coll., Villes, 1998.

[ASC 00] ASCHAN-LEYGONIE C., MATHIAN H., SANDERS L., *et al.*, "A spatial microsimulation of population dynamics in Southern France: a model integrating individual decisions and spatial constraints", in BALLOT G., WEISBUCH G. (eds.), *Applications of Simulation to Social Sciences*, pp. 109–125, 2000.

[ASC 13] ASCHAN-LEYGONIE C., BAUDET S., MATHIAN H., *et al.*, "Gaining a better understanding of respiratory health inequalities among cities: an ecological case study on elderly males in the larger French cities", *International Journal of Health Geographics (IJGH)*, vol. 12, p. 19, 2013.

[AXE 97] AXELROD R., *The Complexity of Cooperation: Agent-Based Models of Competition and Collaboration*, Princeton University Press, 1997.

[AXT 02] AXTELL R.L., EPSTEIN J.M., DEAN J.S., *et al.*, "Population growth and collapse in a multi-agent model of the kayenta anasazi in Long House Valley", *Proceedings of the National Academy of Sciences*, vol. 99, no. 3, pp.7275–7279, 2002.

[AXT 06] AXTELL R., FLORIDA R., Emergent cities: micro-foundations of Zipf's law, Working paper, 2006. Available at http://creativeclass.com/rfcgdb/articles/Emergent_Cities.pdf.

[BAN 01] BANOS A., "A propos de l'analyse spatiale exploratoire des données, *Cybergeo: European Journal of Geography*, Systems, Modeling, Geostatistics", 2001. Available at http://cybergeo.revues.org/4056.

[BAN 13] BANOS A., SANDERS L., "Modéliser et simuler les systèmes spatiaux en géographie", in VARENNE F., SILBERSTEIN M. (eds.), *Modéliser et Simuler – Epistémologies et Pratiques des Modèles et des Simulations*, Ebook des éditions Matériologiques, vol. 1, pp. 833–863, 2013.

[BAT 01] BATTY M., "Polynucleated urban landscapes", *Urban Studies*, vol. 38, no. 4, pp. 635–655, 2001.

[BAT 04] BATTY M., BARROS J., ALVES JUNIOR S., Cities: continuity, transformation, and emergence, Working paper, series 12, CASA, 2004. Available at http://www.bartlett.ucl.ac.uk/casa/pdf/paper72.pdf.

[BAT 05] BATTY M., *Cities and Complexity*, MIT Press, Cambridge, 2005.

[BEN 02] BENENSON I., OMER I., HATNA E., "Entity-based modeling of urban residential dynamics: the case of Yaffo", *Environment and Planning B*, vol. 29, pp. 491–512, 2002.

[BEN 04] BENENSON I., TORRENS P.M., *Geosimulation; Automata-based Modeling of Urban Phenomen,* Wiley, 2004.

[BEN 80] BENZECRI J.-P., *Pratique de l'analyse des Données*, Dunod, 1980.

[BER 11a] BERROIR S., MATHIAN H., SAINT-JULIEN T.H., *et al.*, "The role of mobility in the building of metropolitan polycentrism", in DESROSIERS F., THÉRIAULT M. (eds.), *Modeling Urban Dynamics*, ISTE, London, and John Wiley & Sons, New York, pp. 1–25, 2011.

[BER 11b] BERRY B.J.L., OKULICZ-KOZARYN A., "The city size distribution debate: resolution for US urban regions and megalopolitan areas", *J. Cities*, 2011.

[BER 64] BERRY B., "Approaches to regional analysis: a synthesis", *Annals of the Association of American Geographers*, vol. 54, no. 1, pp. 2–11, 1964.

[BER 67] BERTIN J., *Sémiologie Graphique*, Mouton/Gauthier-Villars, Paris, 1967.

[BIT 04] BITTNER T., DONNELLY M., SMITH B., "Endurants and perdurants in directly depicting ontologies", *AI Communications*, vol. 13, no. 4, pp. 247–258, 2004.

[BOM 04] BOMAN M., HOLM E., "Multi-agent systems, time geography, and microsimulations". in OLSSON M.-O., SJÖSTEDT G. (eds.), *Systems Approaches and Their Application: Examples from Sweden*, Kluwer Academic Publishers, Dordrecht, pp. 95–118, 2004.

[BOU 13] BOULARAN Z., CIESIELSKI H., FRANÇOIS J.-C., *et al.*, "Modéliser la réussite au Brevet des collèges à Paris et sa petite couronne", Communication aux onzièmes rencontres de Théo Quant, Besançon, 2013.

[BRE 99] BRETAGNOLLE A., MATHIAN H., PUMAIN D., *et al.*, "Long-term dynamics of European towns and cities: towards a spatial model of urban growth", *Cybergeo: European Journal of Geography, 11th European Colloquium on Quantitative and Theoretical Geography*, Durham Castle, UK, 3–7 September 1999. Available at http://cybergeo.revues.org/566; DOI: 10.4000/cybergeo.566.

[BRE 05] BRETAGNOLLE A., VERDIER N., "Images d'un réseau en évolution: les routes de poste dans la France pré-industrielle", *M@ppemonde*, no. 79, 2005. Available at http://mappemonde.mgm.fr/num7/articles/art05301.html.

[BRE 11a] BRETAGNOLLE A., GUÉROIS M., MATHIAN H., *et al.*, Harmonie-cités: bases de données harmonisées sur la dynamique et les compétences des villes en réseau selon les régions du monde, Final Report of the ANR and Corpus Tools Research SHS, p. 54, 2011. Available at http://hal.archives-ouvertes.fr/halshs-00688491_v1/.

[BRE 11b] BRETAGNOLLE A., GUÉROIS M., LE NECHET F., *et al.*, "Exploration multi-échelles d'une base de données harmonisée sur les villes d'Europe", *Colloque international du GIS Collège International des Sciences du Territoires, "Fonder les sciences du territoire*", Paris, 23–25 November 2011.

[BRE 12] BRETAGNOLLE A., GUÉROIS M., MATHIAN H., *et al.*, Inception report WPB2 cities, ESPON M4D Multi-dimensional Data Design and Development Inception Report, 2012.

[BRE 13] BRETAGNOLLE A., GUEROIS M., MATHIAN H., *et al.*, Naming U.M.Z: a database now operational for urban studies, ESPON Database 2013 project, final report, 2013. Available at http://www.espon.eu/export/sites/default/Documents/Projects/ScientificPlatform/ESPONDatabase2013/TR_Draft_Final_Report.pdf.

[BRO 98] BROGAARD B., "Fields, objects and dependence relations", in PEUQUET D., SMITH B., BROGAARD B. (eds.), *The Ontology of Fields,* Report of a specialist meeting held under the auspices of the Varenius Project, Bar Harbor, Maine, 11–13 June 1998.

[BUA 13] BUARD E., Dynamiques des interactions espèces – espace, Mise en relation des pratiques de déplacement des populations d'herbivores et de l'évolution de l'occupation du sol dans le parc de Hwange (Zimbabwe), Doctoral Thesis in Geography, University of Paris 1, 2013.

[BUL 05] BULLE N., "Les modèles formels et l'explication en sciences sociales", *L'Année sociologique,* vol. 55, no. 1, pp. 19–34, 2005.

[CAL 00] CALDWELL S., MORRISON R., "Validation of longitudinal microsimulation models: experience with CORSIM and DYNACAN", in MITTON *et al.* (eds.), *Microsimulation in the New Millennium*, Cambridge University Press, Cambridge, 2000.

[CAR 07] CARUSO G., PEETERS D., CAVAILHÈS J., *et al.*, "Spatial configurations in a periurban city. A cellular automata-based microeconomic model", *Regional Science and Urban Economics*, vol. 37, no. 5, pp. 542–567, 2007.

[CHA 87] Groupe CHADULE. (CHAMUSSY H., CHARRE J., DUMOLARD P., *et al.*,), *Initiation Aux Pratiques Statistiques en Géographie*, Colin, Paris, 1987.

[CHA 95] CHARRE J., *Statistique et Territoire*, Reclus-Espace modes d'emploi, Montpellier, 1995.

[CHE 76] CHEN P., "The entity-relationship model – toward a unified view of data", *ACM Transactions on Database Systems*, vol. 1, no. 1, pp. 9–36, 1976.

[CHE 93] CHEYLAN J.-P., LARDON S., "Toward a conceptual model for the analysis of spatio-temporal processes". in FRANK A., CAMPARI I. (eds.), *Spatial Information Theory, COSIT '9 Conference*, Lecture Notes in Computer Science, Springer Verlag, Berlin, vol. 716, pp. 158-176, 1993.

[CHE 95] CHEYLAN J.-P., LARDON S., MATHIAN H., *et al.*, "Les problématiques de l'espace et du temps et les SIG", *Revue Internationale de Géomatique*, no. 4, pp. 287–305, 1995.

[CHE XX] CHEYLAN J.-P., "Time and spatial database, a conceptual application framework", in FRANK A., RAPPER J., CHEYLAN J.-P. (eds.), *Life and Motion of Socio-economic Units*, GISDATA Series, Taylor and Francis, New York/London, no. 8, pp. 35–48.

[CLA 87] CLARKE M., HOLM E., "Micro-simulation methods in spatial analysis and planning", *Human Geography*, Geografiska Annaler, Series B, vol. 69, no. 2, pp. 145–164, 1987.

[CLA 95] CLARAMUNT C., THÉRIAULT M., "Managing time in GIS: an event-oriented approach", in CLIFFORD J., TUZHILIN A. (eds.), *Proceedings of the International Workshop on Temporal Databases: Recent Advances in Temporal Databases*, Springer-Verlag, Zurich, Switzerland, pp. 23–42, 1995.

[CLA 96] CLARKE G. (ed.), *Microsimulation for Urban and Regional Policy Analysis*, Pion, London, 1996.

[CLI 73] CLIFF A.D., ORD J.K., *Spatial Autocorrelation*, Pion, London, 1973.

[COS 11] COSSART E., "Mapping glacier variations at regional scale through equilibrium line altitude interpolation: GIS and statistical application in Massif des Écrins (French Alps)", *Journal of GIS*, vol. 3, no. 3, pp. 232–241, 2011.

[COU 85] COUCLELIS H., "Cellular worlds: a framework for modeling micro – macro dynamics", *Environment and Planning A*, vol. 17, no. 5, pp. 585–596, 1985.

[COU 92] COUCLELIS H., "People manipulate objects (but cultivate fields): beyond the raster-vector debate in GIS", in FRANK A.U., CAMPARI I., ORMENTINI U. (eds.), *Theories and Methods of Spatio-Temporal Reasoning in Geographic Space*, Springer Verlag, Berlin, pp. 65–77, 1992.

[COU 97] COUCLELIS H., "From cellular automata to urban models: new principles for model development and implementation", *Environment and Planning B*, vol. 24, pp. 165–174, 1997.

[COU 98] COUCLELIS H., "Aristotelian spatial dynamics in the age of GIS", in EGENHOFER M., GOLLEDGE R.G. (eds.), *Spatial and Temporal Reasoning in Geographic Information Systems*, Oxford University Press, New York, pp. 109–118, 1998.

[CRE 93] CRESSIE N. *Statistics for Spatial Data*, revised ed., Wiley, New York, 1993.

[DAU 07] DAUDÉ É., LANGLOIS P., "Comparison of three implementations of schelling's spatial segregation model", in PHAN D., AMBLARD F. (eds.), *Agent-Based Modeling and Simulation in the Social and Human Sciences*, GEMAS Studies in Social Analysis, The Bardwell Press, Oxford, pp. 295–326, 2007.

[DAU 13] DAUPHINÉ A., PROVITOLO D., *Risques et Catastrophes: Observer, Spatialiser, Comprendre, Gérer*, 2nd ed., Colin, 2013.

[DAV 01] DAVIDSSON P., "Multi agent based simulation: Beyond social simulation", in MOSS S., DAVIDSSON P. (eds.), *Multi Agent Based Simulation*, Springer Verlag, Berlin Heidelberg LNCS series, vol. 1979, pp. 97–107, 2001.

[DEB 04] DEBARBIEUX B., "De l'objet spatial à l'effet géographique", in DEBARBIEUX B., FOURNY M.-C. (eds.), *L'effet Géographique. Construction Sociale, Appréhension Cognitive et Configuration Matérielle des Objets Géographiques*, CNRS – Maison des Sciences de l'Homme-Alpes, Grenoble, 2004.

[DOU 08] DOUVINET J., DELAHAYE D., LANGLOIS L., "Modélisation de la dynamique potentielle d'un bassin versant et mesure de son efficacité structurelle", *Cybergeo*, no. 412, 2008.

[DRE 04] DRENNAN R., PETERSON C., "Comparing arhaeological settlement systems with rank-size graphs: a measure of shape and statistical confidence", *Journal of Archaeological Science*, vol. 31, no. 5, pp. 533–549, 2004.

[DUB 03] DUBOS-PAILLARD E., GUERMOND Y., LANGLOIS P., "Analyse de l'évolution urbaine par automate cellulaire. Le modèle SpaCelle", *L'Espace Géographique*, 4, vol. 32, 2003.

[DUB 14] DUBÉ J., LEGROS D., *Spatial Econometrics using Microdata*, ISTE, London, and John Wiley & Sons, New York, 2014.

[DUM 11] DUMOLARD P., *Données Géographiques. Analyse Statistique Multivariée*, Hermès Science, Lavoisier, 2011.

[DUR 90] DURAND-DASTÈS F., "La mémoire de Gaïa", *Proceedings of Geopoint 1990: Histoire, Temps et Espace*, Avignon, 1990.

[DUR 92] DURAND-DASTES F., "Le particulier et le général en géographie", in AUDIGIER F., BAILLAT G. (eds.), *Analyser et Gérer les Situations d'enseignement- Apprentissage. La Notion de Chaos. Quelques Réflexions, INRP Actes du 6th colloque de didactique de l'histoire, de la géographie et des sciences sociales*, Paris, pp. 311–314, 1992.

[EDM 05] EDMONDS B., MOSS S., "From KISS to KIDS in multi-agent and multi-agent based simulation", *Lecture Notes in Computer Science,* vol. 3415, pp. 130–144, 2005.

[ENG 97] ENGELEN G., WHITE R., ULJEE I., "Integrating constrained cellular automata models, GIS and decision support tools for urban planning and policy making", in TIMMERMANS H. (ed.), *Decision Support Systems in Urban Planning*, pp. 125–155, 1997.

[ENG 99] ENGELEN G., GEERTMAN S., SMITS P., *et al.*, "Dynamic GIS and strategic physical planning: a practical application", in STILLWELL J., GEERTMAN S., OPENSHAW S. (eds.), *Geographical Information and Planning*, Advances in Spatial Science, Springer, pp. 87–111, 1999.

[EPS 96] EPSTEIN J., AXTELL R., *Growing Artificial Societies: Social Science from the Bottom Up*, MIT Press/Brookings Institution, Cambridge, MA, 1996.

[EPS 99] EPSTEIN J., JOSHUA M., "Agent-based computational models and generative social science", *Complexity*, vol. 4, no. 5, pp. 41–57, 1999.

[EPS 07] EPSTEIN J.M., *Generative Social Science: Studies in Agent-Based Computational Modeling*, Princeton University Press, 2007.

[FAV 98] FAVORY F., MATHIAN H., RAYNAUD C., *et al.*, "Sélection géographique, déterminisme et hasard", *Archaeomedes, Des oppida aux métropoles, Anthropos, coll. Villes*, pp. 151–248, 1998.

[FAV 12] FAVORY F., NUNINGER L., SANDERS L., "Intégration de concepts de géographie et d'archéologie spatiale pour l'étude des systèmes de peuplement", *l'Espace Géographique*, no. 4, pp. 295–309, 2012.

[FER 95] FERBER J., *Les Systèmes Multi-agents. Vers une Intelligence Collective*, InterEditions, Paris, 1995.

[FLE 07] FLEURY A., "Aménager les rues pour changer la ville", in SAINT-JULIEN T., LE GOIX R. (eds.), *La Métropole Parisienne. Centralités, Inégalités, Proximités*, Paris, Belin, pp. 253–276, 2007.

[FOR 02] FOTHERINGHAM S.A, BRUNSDON C., CHARLTON M, *Geographically Weighted Regression: The Analysis of Spatially Varying Relationships*, John Wiley & Sons, Chichester, UK, 2002.

[FOT 00] FOTHERINGHAM S., BRUNDSON C., CHARLTON M., *Quantitative Geography: Perspectives on Spatial Data Analysis*, SAGE, 2000.

[FRA 08] FRANCOIS J.C., POUPEAU F., "Les déterminants socio-spatiaux du placement scolaire. Essai de modélisation statistique appliquée aux collèges parisiens", *Revue française de sociologie*, vol. 49, no. 1, pp. 93–126, 2008.

[FRA 14a] FRANÇOIS J.C., CIESIELSKI H., BOULARAN Z., *et al.*, "Mobilité géographique et réussite au collège dans l'agglomération parisienne", in COURTY G. (ed.), "*Les mobilités dans le système scolaire*", Presses de Sciences Po, 2014.

[FRA 14b] FRANÇOIS J-C., MATHIAN H., SANDERS L., *et al.*, "La structuration sociale de l'espace scolaire: une ontologie multipoints de vue, intégrée mais non réductrice", in PHAN D. (ed.), *Ontologies et Modélisation par SMA en SHS*, Hermes-Lavoisier, London/Paris, 2014.

[GAL 01] GALTON A., "A formal theory of objects and fields, spatial information theory", *Lecture Notes in Computer Science*, vol. 2205, pp. 458–473, 2001.

[GAL 03] GALTON A., "On the ontological status of geographical boundaries", in DUCKHAM M., GOODCHILD M.F., WORBOYS M.F. (eds.), *Foundations of Geographic Information Science*, 2003.

[GAL 04] GALTON A., "Fields and objects in space, time, and space-time", *Spatial Cognition and Computation*, vol. 4, no. 1, pp. 39–68, 2004.

[GAL 09] GALTON A., MIZOGUCHI R., 2009, "The water falls but the waterfall does not fall: new perspectives on objects, processes and events", *Journal Applied Ontology*, vol. 4, no. 2, pp. 71–107, 2009.

[GAR 12] GARMY P., *Villes Réseaux et Systèmes de Villes*, Errance, Actes Sud, 2012.

[GAR 70] GARDNER M., "Mathematical games. The fantastic combinations of John Conway's new solitaire game 'life'", *Scientific American*, no. 223, pp. 120–123, 1970.

[GIL 99] GILBERT N., TROITZSCH K.G., *Simulation for the Social Scientist*, 2nd ed., Open University Press, Milton Keynes, 2005.

[GIR 09] GIRAUT F., VACCHIANI-MARCUZZO C., *Territories and Urbanization, Atlas and geo-historical information system (DYSTURB)*, Paris, IRD Editions, 2009.

[GOU 95] GOURIEROUX C., MONFORT A., *Séries Temporelles et Modèles Dynamiques*, Economica, 1995.

[GRE 04] GRENON P., SMITH B., "SNAP and SPAN: towards dynamic spatial ontology", *Spatial Cognition and Computation*, vol.4, no. 1, pp. 69–103, 2004.

[GRI 11] GRIFFIN A., "Emergence of fusion/fission cycling and self-organised criticality from a simulation model of early complex polities", *Journal of Archaeological Science*, vol. 38, pp. 873–883, 2011.

[GRI 91] GRIFFITH D., AMRHEIN C., *Statistical Analysis for Geographers*, Prentice Hall, Englewood Cliffs, NJ, 1991.

[GRI 10] GRIFFITH D.A., "Modeling spatio-temporal relationships: retrospect and prospect", *Journal of Geographical Systems*, vol. 12, no. 2, pp. 111–123, 2010.

[GRU 93] GRUBER T., "A translation approach to portable ontology specifications", *Knowledge Acquisition*, vol. 5, no. 2, pp. 199–220, 1993.

[GUE 95] GUERIN-PACE F., "Rank-size distribution and the process of urban growth", *Urban Studies*, vol. 32, pp. 551–562, 1995.

[GUE 12] GUEROIS M., BRETAGNOLLE A., GIRAUD T., *et al.*, "A new database for the cities of Europe? Urban Morphological Zones (CLC2000) confronted to three national databases of urban agglomerations (Denmark, France Sweden)", *Environment and Planning B*, vol. 39, no. 3, pp. 439–458, 2012.

[GUM 03] GUMERMAN G.J., SWEDLUND A.C., DEAN J.S., *et al.*, "The evolution of social behavior in the prehistoric american southwest", *Artificial Life*, vol. 9, pp. 435–444, 2003.

[GUO 09] GUO D., MENNIS J., "Spatial data mining and geographic knowledge discovery–an introduction", *Computers, Environment and Urban Systems*, vol. 33, pp. 403–408, 2009.

[HAA 89] HAAG G., *Dynamic Decision Theory: Applications to Urban and Regional Topics*, Kluwer Academic Publishers, Dordrecht, 1989.

[HAG 53] HÄGERSTRAND T., "Innovationsförloppet ur korologisk synpunkt", *Meddelanden från Lunds universitets geografiska institution*, vol. 25, 1953. (English translation by Allan Pred, *Innovation Diffusion as a Spatial Process*, University of Chicago Press, 1967).

[HAG 65] HAGGETT P., *Locational Analysis in Human Geography*, Arnold, London, 1965.

[HAG 70] HÄGERSTRAND T., "What about people in regional science", *Paper of the Regional Science Association*, vol. 24, no. 1, pp. 6–21, 1970.

[HAW 04] HAWLEY K., "Temporal parts", in ZALTA E.N. (ed.), *The Stanford Encyclopedia of Philosophy*, 2004.

[HEC 13] HECKBERT S., "MayaSim: an agent-based model of the ancient maya social-ecological system", *Journal of Artificial Societies and Social Simulation*, vol. 16, no. 4, p. 11, 2013. Available at http://jasss.soc.surrey.ac.uk/16/4/11.html.

[HEP 12] HEPPENSTALL A.J., CROOKS A.T., SEE L.M., *et al.*, (eds.), *Agent-Based Models of Geographical Systems*, Springer, 2012.

[HOL 89] HOLM E., MAKILA K., OBERG S., "Tidsgeografisk *handlingsteori*; att bilda betingade biografier" (A theory on behavior based on "time-geography"; building coherent individual biographies), GERUM, Report no. 8, Umea, 1989.

[HOL 00] HOLM E., URBAN L., MALMBERG G., "Dynamic microsimulation", in FOTHERINGHAM S., WEGENER M. (eds.), *Spatial Models and GIS: New Potential and New Models*, GISDATA 7, Taylor & Francis, pp. 143–165, 2000.

[HOL 02] HOLM E., HOLME K., MÄKILÄ K., *et al.*, "The SVERIGE spatial microsimulation model – content, validation, and example applications", *Gerum Kulturgeografi 4*, Umeå University, 2002.

[HOL 07a] HOLM E., LINDGREN U., HÄGGSTRÖM LUNDEVALLER E., *et al.*, "SVERIGE", in GUPTA A., HARDING A. (eds.), *Modeling our Future, Population Ageing Health and Aged Care, International Symposia in Economic Theory and Econometrics*, Elsevier, vol. 16, 2007.

[HOL 07b] HOLM E., SANDERS L., "Spatial microsimulation models", in SANDERS L. (ed.), *Models in Spatial Analysis*, ISTE, London, 2007.

[HOR 98] HORNSBY K., EGENHOFER M.J., "Identity based change operations for composite objects", in POIKER T., CHRISMAN N. (eds.), *8th International Symposium on Spatial Data Handling*, Vancouver, Canada, pp. 202–213, July 1998.

[HOR 00] HORNSBY K., EGENHOFER M.J., "Identity-based change: a foundation for spatio-temporal knowledge representation", *International Journal of Geographical Information Science*, vol. 14, no. 3, 2000.

[IRV 01] IRWIN E.G., GEOGHEGAN J., "Theory, data, methods: developing spatially explicit economic models of land use change", *Agriculture, Ecosystems and Environment*, vol. 85, pp. 7–23, 2001.

[JAN 09] JANSSEN M., "Understanding artificial anasazi", *Journal of Artificial Societies and Social Simulation,* vol. 12, no. 4, p. 13, 2009.

[KLE 07] KLEIN E., *Le Facteur Temps ne Sonne Jamais Deux Fois*, Flammarion, 2007.

[KOH 00] KOHLER T.A., GUMERMAN G.J. (eds.), *Dynamics in Human and Primate Societies: Agent-Based Modeling of Social and Spatial Processes*, Oxford University Press, New York, NY, 2000.

[KOH 07] KOHLER T.A., VAN DER LEEUW S.E. (eds.), *The Model-based Archaeology of Socionatural Systems*, School for Advanced Research Press, 2007.

[KOH 08] KOHLER T.A., VARIEN M.D., WRIGHT A.M., *et al.*, "Mesa verde migrations", *American Scientist*, vol. 96, 2008.

[LAJ 07] LAJOIE G, HAGEN-ZANKER A., "La simulation de l'étalement urbain à La Réunion: apport de l'automate cellulaire Metronamica pour la prospective territoriale", *Cybergeo: European Journal of Geography*, 2007.

[LAN 88] LANGRAN G., CHRISMAN N., "A framework for temporal geographic information", *Cartographica*, vol. 25, no. 3, pp. 1–14, 1988.

[LAU 86] LAULAJAINEN R., GADDE L.-E., "Locational avoidance: a case study of three Swedish retail chains", *Regional Studies*, vol. 20, no. 2, pp. 131–140, 1986.

[LEB 06] LEBART L., MORINEAU A., PIRON M., *Statistique Exploratoire Multidimensionnelle; Visualisation et Inférence en Fouille de Données*, Dunod, 2006.

[LEF 12] LEFEBVRE B., RODIER X., SALIGNY L., "La modélisation de l'information spatio-temporelle", in DE BEAUNE S.A., FRANCFORT H.-P. (eds.), *L'archéologie à Découvert*, CNRS, Paris, pp. 249–258, 2012.

[LEN 09] LENCLUD G., 2009, "Les cultures humaines et le bateau de Thésée. Le problème de l'identité des cultures", in LABORDE D. (ed.), *Désirs d'histoire*, L'Harmattan, Paris, pp. 221–248, 2009.

[LEW 88] LEWIS D., *On the Plurality of Worlds*, Blackwell, Oxford, 1986.

[LIV 09] LIVET P., NEF F., *Les êtres Sociaux, Processus et Virtualité*, Hermann Philosophie, 2009.

[LIV 10] LIVET P., MULLER J-P., PHAN D., *et al.*, "Ontology, a mediator for agent based modeling in social science", *Journal of Artificial Society and Social Simulation*, vol. 13, no. 1, 2010.

[LIV 14] LIVET P., SANDERS L., "Le 'test ontologique': un outil de médiation pour la modélisation agent" in PHAN D. (ed.), *Ontologies et Modélisation par SMA en SHS*, Hermes/Lavoisier, London/Paris, 2014.

[LON 96] LONGLEY P.A., BATTY M. (eds.), *Spatial Analysis: Modeling in a GIS Environment*, GeoInformation International, Cambridge, 1996.

[MAC 09] MACY M.W., FLACHE A., "Social dynamics from the bottom up: agent-based models of social interaction", in HEDSTRÖM P., BEARMAN P. (eds.), *The Oxford Handbook of Analytical Sociology*, Oxford University Press, pp. 245–268, 2009.

[MAN 07] MANZO G., "Progrès et urgence de la modélisation en sociologie. Du concept de modèle générateur et de sa mise en œuvre", *L'Année Sociologique*, vol. 57, no. 1, pp. 13–61, 2007.

[MAN 09] MANZO G., *La Spirale des Inégalités; Choix Scolaires en France et en Italie au XXe siècle*, PUPS, Paris, 2009.

[MAR 91] MARCH J., "Exploration and exploitation in organizational learning", *Organization Science*, vol. 2, pp. 71–87, 1991.

[MAT 00] MATHIAN H., SANDERS L., MIKULA D., "Modeling the dynamics of spatial systems within a GIS: problems and perspectives", in FOTHERINGHAM A.S., WEGENER M. (eds.), *Spatial Models in GIS*, GISDATA 7, Taylor & Francis, London, pp. 203–220, 2000.

[MAT 06] MATHIAN H., SANDERS L., "Scientific approach of the MAUP?", in GRASLAND C., MADELIN M. (eds.), *The Modifiable Areal Unit Problem, Final Report of ESPON 3.4.3, ESPON 2000–2006 Program, Partly Financed Through the INTERREG III ESPON 2006 Program*, pp. 1–33, 2006

[MAT 07] MATHIAN H., PIRON M., "Geographical scales and multidimensional statistical methods", in SANDERS L. (ed.), *Models in Spatial Analysis*, ISTE, pp. 62–103, 2007.

[MAT 08] MATHIAN H., SANDERS L., "Expérimenter sur le futur des villes européennes avec un modèle multi-agents", *15e Journées de Rochebrune, Rencontres interdisciplinaires sur les systèmes complexes naturels et artificiels*: "*Expérimentation et systèmes complexes*", ENST 2008 S 001, Paris, 2008.

[MCK 02] MCKINNON N., "The endurance/perdurance distinction", *Australasian Journal of Philosophy*, vol. 80, no. 3, pp. 288–306, 2002.

[MEN 00] MENNIS J.L., PEUQUET D.J., QIAN L., "A conceptual framework for incorporating cognitive principles into geographical database representation", *International Journal of Geographical Information Science*, vol. 14, no. 6, pp. 501–520, 2000.

[MIL 09] MILLER H.J., HAN J., "Geographic data mining and knowledge discovery: an overview", in MILLER H.J., HAN J. (eds.), *Geographic Data Mining and Knowledge Discovery: An Overview*, 2nd ed., CRC Press, 2009.

[MIN 65] MINSKY M.L., "Matter, mind and models", *Proceedings of the International Federation of Information Processing Congress*, vol. 1, pp. 45–49, 1965.

[MOE 03] MOECKEL R., SPIEKERMANN K., SCHÜRMANN C., *et al.*, "Microsimulation of land use", *International Journal of Urban Sciences*, vol. 7, no. 1, pp. 14–31, 2003.

[MOR 06] MORENCY C., "Etude des méthodes d'analyse spatiale et illustration à l'aide de microdonnées urbaines de la grande région de Montréal", *Les Cahiers Scientifiques du Transport*, no. 49, pp. 77–102, 2006.

[MOR 09] MORENO D., BADARIOTTI D., BANOS A., "Graph based automata for urban modeling", *in* BAVAUD F., MAGER C. (eds.), *Handbook of Theoretical and Quantitative Geography*, FGSE, University of Lausanne, Switzerland, pp. 261–309, 2009.

[MOR 93] MORICONI-EBRARD F., *L'urbanisation du Monde depuis 1950*, Anthropos, Paris, 1993.

[NUN 06] NUNINGER L., SANDERS L., FAVORY F., *et al.*, "La modélisation des réseaux d'habitat: trois expériences", *M@ppemonde*, vol. 83, no. 3, p. 28, 2006.

[OLT 12] OLTEANU-RAIMOND A-M., COURONNE T., FEN-CHONG J., *et al.*, "Le Paris des visiteurs étrangers, qu'en disent les téléphones mobiles? Inférence des pratiques spatiales et fréquentations des sites touristiques en Ile-de-France", *Revue Internationale de Géomatique*, 2012.

[OPE 81] OPENSHAW S., "Le problème de l'agrégation spatiale en géographie", *Espace Géographique*, vol. 10, no. 1, pp. 15–24, 1981.

[OPE 84] OPENSHAW S., *The Modifiable Areal Unit Problem*, CAT- MOG 38, GeoBooks, Norwich, England, 1984.

[ORC 57] ORCUTT G., "A new type of socio-economic system", *Review of Economics and Statistics*, no. 58, pp. 773–797, 1957.

[OSU 05] O'SULLIVAN D., "Geographical information science: time changes everything", *Progress in Human Geography*, vol. 29, no. 6, pp.749–756, 2005.

[OSU 10] O'SULLIVAN D., UNWIN D., *Geographic Information Analysis*, 2nd ed., John Wiley & Sons, 2010.

[PAR 08] PARISI D., ANTINUCCI F., NATALE F., *et al.*, "Simulating the expansion of farming and the differentiation of european languages", in LAKS B. (ed.), *Origin and Evolution of Languages Approaches, Models, Paradigms*, Equinox Publishing, 2008.

[PAU 10] PAULUS F., VACCHIANI-MARCUZZO C., "Innovative activities and city size: a cross-country analysis (France, USA)", *Seminar in Center for Urban and Regional Analysis*, OSU, Columbus, 5 November 2010.

[PEA 80] PEARSON C., "Rank-size distributions and the analysis of prehistoric settlement systems", *Journal of Anthropological Research*, vol. 36, no. 4, pp. 453–462, 1980.

[PEU 84] PEUQUET D.J. "A conceptual framework and comparison of spatial data models", *Cartographica*, vol. 21, no. 4, pp. 66–113, 1984.

[PEU 88] PEUQUET D.J., "Representations of geographic space: toward a conceptual synthesis", *Annals of the Association of American Geographers*, vol. 78, pp. 375–394, 1988.

[PEU 94] PEUQUET D.J., "A conceptual framework for the representation of temporal, dynamics in geographic information systems", *Annals of the Association of American, Geographers*, vol. 84, no. 3, pp. 441–461.

[PEU 95] PEUQUET D.J., DUAN N., "An event-based spatiotemporal data model (ESTDM) for temporal analysis of geographical data", *International Journal of Geographical Information System*, vol. 9, no. 1, pp. 7–24, 1995.

[PEU 98] PEUQUET D.J., SMITH B., BROGAARD B. (eds.), The ontology of fields, Report of a specialist meeting held under the auspices of the Varenius Project, Maine, Bar Harbor, 11–13 June 1998.

[PEU 02] PEUQUET D.J., *Representations of Space and Time*, Guilford, New York, 2002.

[PHA 07] PHAN D., AMBLARD F. (eds.), *Agent-based Modeling and Simulation in the Social and Human Sciences*, The Bardwell Press, Oxford, 2007.

[PHA 14] PHAN D., (ed.), *Ontologies et Modelisation par SMA en SHS*, Hermès-Lavoisier, Paris, 2014.

[PLE 98] PLEWE B., Data modeling, GIS integration, vagueness in PEUQUET D., SMITH B., BROGAARD B. (eds.), The Ontology of Fields, Report of a specialist meeting held under the auspices of the Varenius Project, Maine, Bar Harbor, 11–13 June 1998.

[PLU 11] PLUMEJEAUD C., Modèles et méthodes pour l'information spatio-temporelle évolutive, Doctoral Thesis, University of Grenoble, 2011.

[PHI 97] PHIPPS M., LANGLOIS A., "Spatial dynamics, cellular automata, and parallel processing computers", *Environment and Planning B : Planning and Design*, vol. 24, no. 2, 1997, pp. 687–705, 1997.

[POR 97] PORTUGALI J., BENENSON I., OMER I., "Spatial cognitive dissonance and sociospatial emergence in self-organizing", *Environment and Planning*, vol. 24, no. 2, pp. 263–285, 1997.

[PUM 97] PUMAIN D., SAINT-JULIEN T., *L'analyse Spatiale*, Localisations dans l'espace, Cursus, Colin, vol. 1, p. 167, 1997.

[PUM 06] PUMAIN D., "Alternative explanations of hierarchical differentiation in urban systems", in PUMAIN D. (ed.), *Hierarchy in Natural and Social Sciences*, Methodos series 3, Springer, pp. 169–222, 2006.

[PUM 09] PUMAIN D., SANDERS L., BRETAGNOLLE A., *et al.*, "The future of urban systems: exploratory models" in VAN DER LEEUW L., WEST P. (eds.), *Complexity Perspectives on Innovation and Social Change*, Methodos series, Springer, pp. 331–359, 2009.

[RAP 95] RAPER J., LIVINGSTONE D., "Development of a geomorphologic spatial model using object-oriented design", *International Journal of Geographical Information Systems*, vol. 9, no. 4, pp. 359–384, 1995.

[REN 96] RENOLEN A., "History graphs: conceptual modeling of spatiotemporal data", *International Cartographic Association, Proceedings of GIS Frontiers in Business and Science*, Brno, Czech Republic, vol. 2, 1996.

[REN 00] RENOLEN A., "Modeling the real world: conceptual modeling in spatiotemporal information system design", *Transactions in GIS*, vol. 4, no. 1, pp. 23–42, 2000.

[REP 04] REPHANN T.J., HOLM E., 2004, "Economic-demographic effects of immigration: results from a dynamic spatial microsimulation model", *Interregional Regional Science Review*, vol. 27, no. 4, pp. 379–410, 2004.

[RIP 05] RIPLEY B.D., *Spatial Statistics*, John Wiley & Sons, 2005.

[ROD 10] RODIER X., SALIGNY L., "Modélisation des objets historiques selon la fonction, l'espace et le temps pour l'étude des dynamiques urbaines dans la longue durée", *Cybergeo: European Journal of Geography*, Systèmes, Modélisation, Géostatistiques, 2010. Available at http://cybergeo.revues.org/index23175.html.

[SAN 89] SANDERS L., *L'analyse Statistique des Données en Géographie*, Alidade-RECLUS, Montpellier, p. 268, 1989.

[SAN 97a] SANDERS L., "Durability of settlements systems: a long term perspective", *Cybergeo: European Journal of Geography*, no. 31, 1997. Available at http://cybergeo.revues.org/1547.

[SAN 97b] SANDERS L., PUMAIN D., MATHIAN H., *et al.*, "SIMPOP : a multi-agents system for the study of urbanism", *Environment and Planning B*, vol. 24, pp. 287–305, 1997.

[SAN 05] SANDERS L., FAVARO J.-M., MATHIAN H., *et al.*, "Intelligence artificielle et agents collectifs: le modèle EUROSIM", *Cybergeo: European Journal of Geography*, Sélection des meilleurs articles de SAGEO, 2005.

[SAN 10] SANDERS L.,2010, "Les villes comme agents : simulation des futurs possibles du système urbain européen", *NPSS (Nouvelles Perspectives en Sciences Sociales)*, vol. 5, no. 2, pp. 153–180, May 2010.

[SAN 12] SANDERS L., "Regards croisés de géographes, économistes et archéologues sur la hiérarchie des systèmes de peuplement: de l'empirie aux systèmes complexes", in SCHAFFAR A. (ed.), spécial *Région et Développement*, no. 36 "*Hiérarchies et croissances urbaines - Nouveaux regards sur les lois de Zipf et de Gibrat pour les villes*", pp. 127–146, 2012. Available at http://region-developpement.univ-tln.fr/fr/pdf/R36/7_Sanders.pdf.

[SCH 71] SCHELLING T.S., "Dynamic models of segregation", *Journal of Mathematical Sociology*, vol. 1, pp. 143–186, 1971.

[SCH 09] SCHAFFAR A., "La loi de Zipf dans la science régionale: entre anciennes controversies et nouvelles perspectives", *Cybergeo*, Article 450, 2009.

[SCH 13] SCHMITT C., PUMAIN D., "Modélographie multi-agents de la simulation des interactions sociétés-environnement et de l'émergence des villes", *Cybergeo*, Article 643, 2013.

[SER 01] SERNEELS S., LAMBIN E.F., "Proximate causes of land-use change in Narok district, Kenya: a spatial statistical model", *Agriculture, Ecosystems and Environment*, vol. 85, pp. 65–81, 2001.

[SHE 04] SHEEREN D., MUSTIÈRE S., ZUCKER J-D., "Consistency assessment between multiple representations in geographical databases: a specification-based approach", in FISHER P. (ed.), *Developments in Spatial Data Handling (SDH '04)*, Springer, pp. 617–628, 2004.

[SMI 94] SMITH B., "Fiat objects", in GUARINO N., VIEU L., PRIBBENOW S. (eds.), *Parts and Wholes: Conceptual Part-Whole Relations and Formal Mereology, 11th European Conference on Artificial Intelligence*, Amsterdam, 8 August 1994.

[SMI 98] SMITH B., "An introduction to ontology", in PEUQUET D., SMITH B., BROGAARD B. (eds.), The Ontology of Fields, Report of a specialist meeting held under the auspices of the Varenius Project, *Panel on Computational Implementations of Geographic Concepts*, 1998.

[SMI 00] SMITH B., VARZI A.C., "Fiat and bona fide boundaries", *Philosophy and Phenomenological Research*, vol. 60, no. 2, pp. 401–420, 2000.

[SMI 03] SMITH B., "Ontology", in FLORIDI L. (ed.), *Blackwell Guide to the Philosophy of Computing and Information*, Blackwell, Oxford, pp. 155–166, 2003.

[SMI 05] SMITH M.E., "City size in late postclassic Mesoamerica", *Journal of Urban History*, vol. 31, no. 4, pp. 403–434, 2005.

[STA 10] STANILOV K., BATTY M., "Exploring the historical determinants of urban growth through cellular automata", *Working Papers Series 157*, CASA. 2010.

[STR 05] STRAUCH D., MOECKEL R., WEGENER M., *et al.*, "Linking transport and land use planning: the microscopic dynamic simulation model ILUMASS", in ATKINSON P.M., FOODY G.M., DARBY S.E., *et al.* (eds.), *Geodynamics*, CRC Press, Boca Raton, FL, pp. 295–311, 2005.

[TAY 77] TAYLOR P.J., *Quantitative Methods in Geography. An Introduction to Spatial Analysis*, Waveland Press Inc, Prospect Heights, Illinois, IL, 1977.

[TER 98] TERRIER C., "Zonage de pouvoir, zonage de savoir" in BENOIT J.-M., BENOIT P., PUCCI D. (eds.), *La France Redécoupée, Enquête sur la Quadrature de l'hexagone*, Éditions Belin, 1998.

[TOB 70] TOBLER W.R. "A computer movie simulating urban growth in the Detroit region", Supplement: *Proceedings. International Geographical Union. Commission on Quantitative Methods*Clark, University,, vol. 46, pp. 234–240, June 1970.

[TOB 79] TOBLER W.R., "Cellular geography" in GALE S., OLSSON G. (eds.), *Philosophy in Geography*, D. Reidel, Dordrecht, pp. 379–386, 1979.

[TUK 77] TUKEY J., *Exploratory Data Analysis*, Addison-Wesley, 1977.

[UDE 02] UDEHN L., "The changing face of methodological individualism", *Annual Review of Sociology.*, vol. 28, pp. 479–507, 2002.

[UNW 90] UNWIN A.R., WILLS G., HASLETT J., "REGARD – graphical analysis of regional data", *ASA Proceedings of the Section on Statistical Graphics*, pp. 36–41, 1990.

[UNW 94] UNWIN A., "L'exploration des données sour REGARD", *Mappemonde*, vol. 3, 1994.

[VAC 10] VACCHIANI-MARCUZZO C., PAULUS F., "Innovative activities and city size: a cross-country analysis (France, USA)", *Seminar in Center for Urban and Regional Analysis*, OSU, Columbus, 5 November 2010.

[VAN 05] VAN DELDEN H., ESCUDERO J.C., ULJEE I., *et al.*, "METRONAMICA: A dynamic spatial land use model applied to Vitoria-Gasteiz", *Virtual Seminar of the MILES Project, Centro de Estudios Ambientales*, Vitoria-Gasteiz, 2005.

[VAN 09] VAN VLIET J., WHITE R., DRAGICEVIC S., "Modeling urban growth using a variable grid cellular automaton", *Computers, Environment and Urban Systems*, vol. 1, 2009.

[VAN 10] VAN HAMME H., BRETAGNOLLE A., MATHIAN H., "Animer les villes. Exploration dynamique de bases de données urbaines harmonisées dans le temps long", *Mappemonde*, vol.3, no. 99, 2010.

[WAC 99] WACHOWICZ M., *Object-Oriented Design for Temporal GIS*, Taylor & Francis, London, 1999.

[WAC 02] WACHOWICZ M., "Uncovering spatio-temporal patterns in environmental data", *Water Resources Management International Journal*, Special issue on geocomputation in water resources and environment, vol. 16, no. 6, pp. 469–487, 2002.

[WAD 02] WADDELL P., "UrbanSim: modeling urban development for land use, transportation and environmental planning", *Journal of the American Planning Association*, vol. 68, no. 3, pp. 297–314, 2002.

[WAG 07] WAGNER P., WEGENER M., "Urban land use, transport and environment models; experiences with an integrated microscopic approach", *disP*, vol. 170, no. 3, pp. 45–56, 2007.

[WEI 06] WEIDLICH W., *Sociodynamics – A Systematic Approach to Mathematical Modeling in the Social Sciences*, CRC Press, 2000. Republication by Dover Publications, 2006.

[WHI 93] WHITE R., ENGELEN G., "Cellular automata and fractal urban form: a cellular modeling approach to the evolution of urban land-use patterns", *Environment and Planning A*, vol. 25, no. 8, pp. 1175–1199, 1993.

[WHI 94a] WHITE R., "Urban systems dynamics and cellular automata: fractal structures between order and chaos", *Chaos, Solitons & Fractals*, vol. 4, no. 4, pp. 563–583, April 1994.

[WHI 94b] WHITE R., ENGELEN G., "Cellular dynamics and GIS: modeling spatial complexity", *Geographical Systems*, vol. 1, pp. 237–253, 1994.

[WIL 98] WILLIAMSON P., BIRKIN M., REES P H., "The estimation of population microdata by using data from small area statistics and samples of anonymised records", *Environment and Planning A*, vol. 30, no. 5, pp. 785–816, 1998.

[WOR 94] WORBOYS M.F., "A unified model of spatial and temporal information", *Computer Journal*, vol. 37, no. 1, pp. 26–34, 1994.

[WOR 98] WORBOYS M.F., "A generic model for spatio-bitemporal geographic information", in EGENHOFER M.J., GOLLEDGE R.G. (eds.), *Spatial and Temporal Reasoning in Geographic Information Systems*, Oxford University Press, pp. 25–39, 1998.

[WOR 05] WORBOYS M.F., "Event-oriented approaches to geographic phenomena", *International Journal of Geographical Information Science*, vol. 19, pp. 1–28, 2005.

[WRI 02] WRIGLEY N., *Categorical Data Analysis for Geographers and Environmental Scientists*, The Blackburn Press, p. 412, 2002.

[YOU 02] YOUNG D., "A new space-time computer simulation method for human migration, american anthropologist", *New Series*, vol. 104, no. 1, pp. 138–158, 2002.

[YUA 01] YUAN M., "Representing complex geographic phenomena with both object- and field-like properties", *Cartography and Geographic Information Science*, vol. 28, no. 2, pp. 83–96, 2001.

[YUA 99] YUAN M., "Representing geographic information to enhance GIS support for complex spatiotemporal queries", *Transactions in GIS*, vol. 3, no. 2, pp. 137–160, 1999.

[ZIP 49] ZIPF G.K., *Human Behavior and the Principles of Least Effort*, Addinson-Wesley Press, Cambridge, MA, 1949.

Index

D, E

G, H, I, L, M

O, P, Q

R, S

T

Other titles from

ISTE

in

Geographical Information Systems

2014

DUBÉ Jean, LEGROS Diègo
Spatial Econometrics Using Microdata

HÉNO Raphaële, CHANDELIER Laure
3D Modeling of Buildings: Outstanding Sites

PLANTIN Jean-Christophe
Participatory Mapping

2013

SALLABERRY Christian
Geographical Information Retrieval in Textual Corpora

2012

BUCHER Bénédicte, LE BER Florence
Innovative Software Development in GIS

2011

BANOS Arnaud, THÉVENIN Thomas
Geographical Information and Urban Transport Systems

DAUPHINÉ André
Fractal Geography

2010

BRUNET Roger
Sustainable Geography

CARREGA Pierre
Geographical Information and Climatology

CAUVIN Colette, ESCOBAR Francisco, SERRADJ Aziz
Thematic Cartography – 3-volume series
Thematic Cartography and Transformations – volume 1
Cartography and the Impact of the Quantitative Revolution – volume 2
New Approaches in Thematic Cartography – volume 3

LANGLOIS Patrice
Simulation of Complex Systems in GIS

MATHIS Philippe
Graphs and Networks – 2nd edition

THERIAULT Marius, DES ROSIERS François
Modeling Urban Dynamics

2009

ROCHE Stéphane, CARON Claude
Organizational Facets of GIS

2008

BRUGNOT Gérard
Spatial Management of Risks

FINKE Gerd
Operations Research and Networks

GUERMOND Yves
Modeling Process in Geography

KANEVSKI Michael
Advanced Mapping of Environmental Data

2007

DOBESCH Hartwig, DUMOLARD Pierre, DYRAS Izabela
Spatial Interpolation for Climate Data

SANDERS Lena
Models in Spatial Analysis

2006

CLIQUET Gérard
Geomarketing

DEVILLERS Rodolphe, JEANSOULIN Robert
Fundamentals of Spatial Data Quality

Printed and bound by CPI Group (UK) Ltd, Croydon, CR0 4YY

13/09/2026

14961076-0008